T0192845

Requirements Engineering for Social Sector Software Applications

Varun Gupta

Requirements Engineering for Social Sector Software Applications

Innovating for a Diverse Set of User Needs

Varun Gupta
University of Beira Interior
Covilhã, Portugal

This work was supported by Operação Centro-01-0145-FEDER-000019–C4–Centro de Competências em Cloud Computing, co-financed by the Programa Operacional Regional do Centro (CENTRO 2020), through the Sistema de Apoio à Investigação Científica e Tecnológica–Programas Integrados de IC&DT.

ISBN 978-3-030-83551-4 ISBN 978-3-030-83549-1 (eBook)
https://doi.org/10.1007/978-3-030-83549-1

This Springer imprint is published by the registered company Springer Nature Switzerland AG
The registered company address is: Gewerbestrasse 11, 6330 Cham, Switzerland

Foreword

As a computer scientist and former department chair at the Computer Science and Engineering Department, University of Beira Interior, Covilhã, Portugal, it has been my pleasure to know Dr. Varun Gupta since he joined our Cloud Computing Competence Center (C4) as a research fellow (postdoctoral researcher). Before joining C4, he was also an assistant professor at Amity University Noida, India. It has been very special to watch how much he has grown in her recent career.

Requirements Engineering for Social Sector Software Applications: Innovating for a Diverse Set of User Needs, by Varun Gupta, addresses requirements engineering for the social sector (e.g., education, health, non-governmental organisations and charitable trusts) through a multidisciplinary approach. Recall that requirements engineering is nuclear in software engineering. Therefore, this book is handy for those aiming to develop social sector applications in an integrated, multivariate manner, from the software engineers' skills to the industry and social sector requirements, without missing out on the end-user expectations. Interestingly, after introducing the theme of requirements engineering in the social sector context, the author immediately tackles gamification and crowdsourcing techniques in requirements engineering, thus conveying the will of bringing recent approaches from academia to the arena of innovation in the social sector.

His experience in the education sector as an academic, researcher and software developer reflects itself on the build-up of this book, providing a broad and knowledgeable vision about what is happening in the cutting edge of requirements engineering for the social sector, either in academia or in enterprise practice. Therefore, on the verge of digital transformation worldwide, the book is handy for those who intend to digitise social sector institutions, enterprises and start-ups.

Abel J.P. Gomes
Department of Computer Science
University of Beira Interior
Covilhã, Portugal

Contents

Chapter 1
Introduction

Abstract This chapter provides introductory information about the social sector, social innovation and technologically infused innovation. The role of Requirements Engineering (RE) is crucial to have a good quality software product for technologically infused innovation for social good. There is, however, a need to consider the beneficiary of the innovation (i.e., citizen) in software development process as the success of the software depends on his successful adoption of the technology only.

Keywords Crowdsourcing · Gamification · Requirements engineering · Social sector requirements engineering · Social entrepreneurs

1.1 Introduction

Social sector (also called third sector) is that part of the economy that includes the activities for social good, i.e., providing benefits to the society by addressing the social problems like poverty, health issues, lack of education, hygiene, hunger, etc. The society has several problems and efforts are being made by the government as well as non-government organisations (like charitable trusts, social entrepreneurial start-ups, etc.) to contribute in the best possible way for social innovation. Social innovation refers to the improvement of the social sector by bringing new solutions in the form of products and services in the society.

In other words, these products or services have social impact, i.e., they provide benefits to the society. One possible way to bring social innovation is through technologically infused innovation, i.e., bringing technologically improved products or services in the society to provide solutions to societal problems. The examples could include a blockchain system to make financial benefits transparent, an online system to register for unemployment benefits, an online system to book hospital rooms during a pandemic, an online system to provide food to vulnerable groups, etc. These examples provide some meaningful contextual information about the technological infused social innovation:

V. Gupta, *Requirements Engineering for Social Sector Software Applications*,
https://doi.org/10.1007/978-3-030-83549-1_1

- The user base (or beneficiary) of the technological software system is too diverse in terms of their demographics, behaviour, etc. They could be vulnerable groups with less access to the technological competencies, highly technology Savvy group, etc. This factor influences the way the software developers will involve them during software development activities.
- There could be numerous solutions for the same societal problems. Also, numerous problems may exist, and it will be challenging to identify the most promising one where solutions could generate much greater value.

Social innovation has been the focus of government these days; innovation that is absorbed in the society through technological solutions. Technological solutions achieving social innovation need software engineers to explore the problem domain to better understand the social challenges; to identify the value proposition that best addresses the social needs.

The mapping between the technological solutions and actual customer needs (which happens to be government institutions and/or citizens) is the responsibility of the software engineers; the task that is hard to perform due to the interdisciplinary nature of the solutions needed.

The Requirements Engineering (RE) will be challenging to be performed with the wide array of users of social applications. However, if the sponsor of the social application, for instance the federal government, the challenge is to avoid overlooking the customer's actual needs when the federal government represents the possible users in the software development process.

1.2 Requirements Engineering in General

Requirements engineering is the software engineering sub-activity that tries to identify the real needs of customers which are then documented as requirements after a careful analysis, prioritisation and validation [1]. This activity requires continuous interactions with the users either through face-to-face interactions (for instance, through interviews) or through techniques like crowdsourced based RE [2–6]. The formal approach comes under the traditional RE category while latter are advanced RE techniques that best exploits the technology to involve user crowds, thereby leading to larger requirement lists.

In crowd based RE, the RE tasks are outsourced to the crowds. The crowd then provide their perspectives about future requirements of the software, which are then refined further by their continuous social interactions in form of commenting, voting, branching, etc. The crowd based RE can be supported through specialised platforms designed for facilitating the RE or through existing platforms like social media sites like Facebook. The feedback is then analysed by natural language processing techniques and machine learning methods to automatically identify the software requirements along with the rankings.

Crowd based RE have the benefit to involve a larger number of users which result in accurate identification of the requirement set. Contrary, the accuracy had been still uncertain as the efficient analysis techniques with ability to analyse the noisy and larger number of feedbacks is yet to be reported in the literature.

The global reach of such techniques has wider applicability in the social sector which is geographically dispersed throughout the nation boundary. However, there could be multiple limitations when it comes to the use of technology to connect with the masses.

1.3 Social Sector Applications: Divide Between Users, Funders and Customers

The users of the social sector application could be the Government or direct citizens. However, irrespective of who is going to be using the application, the ultimate beneficiary is the citizen. For instance, the federal government could come up with an application to provide unemployment benefits to the citizens affected by coronavirus. If the government is the user, then this application will directly identify the affected parties (for instance, using tax information) otherwise citizens as the users have to apply to get the benefit. In both cases, the user is the direct beneficiary. Why this classification matters, bases its reasons on the ability of the application to have perfect match with market needs, i.e., ability to provide expected benefits to the citizens. This makes it important to explore the problem domain (social problems) and identify the solutions to best address the social needs. This is only possible if there exist no knowledge gaps between the citizen expectations and the requirements engineering team designing the application.

The ability to foster social innovation through products or services with social impact could be a game changer for federal governments in claiming re-elections as well as for economic growth of the nation. The social projects are funded either by governments or by donations as the objective is to have social impact rather than earning profits to the shareholders. However, the social entrepreneurs could have initial investment (called equity) to launch the business, with further operations scaled with scaling donations.

The social entrepreneurs launch the social enterprises (or start-ups) to solve the social problem (business idea) based on their motivation to contribute to the noble cause. The source of the problem could be their professional experience in social sector or their observations based on continuous interactions with the societal issues. The federal government could identify the areas for social innovation based on the rich information they have about the different social sector performances. The reason is that federal government have all the resources that keep on collecting the performance related aspects on continuous bases and experts who recommend which area needs improvement and how. For instance, visiting social institution for claiming the food benefits could result in increased workload across all institutions and heavy demand for the human resources. To make this process easier to citizens and reduce the work pressure, the federal government could come up with the online application which automated this entire value chain.

However, the software application is never adopted by the citizens in one go. In fact, it takes long time for being adopted because of following reasons:

- Hesitation to change and use new ways of doing the things.
- Complex technology and lack of competencies to handle it.
- Application not actually addressing the main pain points of the citizens.

It is essential to consider the citizen diversities in mind while designing the software applications. The continuous updates to the applications are costly to be undertaken and implemented. It is important to decide the software features considering the balance between the funders, users and customers with the focus on providing actual benefits to the citizens. It is also important not to make educated guesses about the citizen needs but rather the hypothesis should be validated with diverse users before starting the development process.

1.4 Requirements Engineering for Social Sector Applications (Social Sector Requirements Engineering)

The RE activity is a co-creation activity that needs continuous support of the users as only by their involvement, requirement analysts will be able to explore problem domain and develop the product with ability to meet the needs. As mentioned in preceding sections and in coming chapters, crowdsourcing has been one way of involving diverse users in the RE process and gamification is a mechanism to foster their motivation. However, to use crowdsourcing, the suitable platform needs to be established to allow them to specify their requirements and take part in social interactions to refine requirements specified by the crowd, leading to the requirement ranking. However, the basic assumption is that the crowd is having competencies to easily use the platform and take part in the RE process. However, even though they have competencies, literature reports that user feedback, which is a rich source of software requirements [7], contains a lot of noises (like jargon) which make their analysis quite difficult [8]. Furthermore, analysing the crowd feedback using natural language processing techniques and machine learning is harder due to non-availability of the advanced and reliable methods [9].

It is interesting to see if existing advanced RE techniques can be feasible to be executed in a social sector context? The reason is grounded in the diversity and lack of accessibility of the citizens that comprise the user crowd of the potential application. The results as disseminated in this book indicate that the existing techniques have limited applicability in social context due to diversity in users and new techniques are required to address the challenges unique to them, called social sector requirements engineering (SSRE).

1.5 Book Organisation

This book is organised as follows, the tertiary study of crowdsourcing and gamification research in the social sector is conducted in Chap. 2 to identify the support that literature can provide in social sector RE. Chapter 3 through multiple case studies explores the area of RE in the social sector (SSRE) and Chap. 4 provides a solution for it. The implications for various stakeholders of the innovation ecosystem are then presented in Chap. 5.

1.6 Conclusion and Future Work

The introductory concepts are presented in this chapter. The large user base of social application is challenging to be involved in designing the technological solutions for social innovation. The federal governments, social entrepreneurs or other social sector innovation partner should focus on more interdisciplinary nature of the social innovation solutions.

References

1. V. Gupta, J.M. Fernandez-Crehuet, T. Hanne, R. Telesko, Requirements engineering in software startups: a systematic mapping study. Appl. Sci. **10**, 6125 (2020). https://doi.org/10.3390/app10176125
2. P.K. Murukannaiah, N. Ajmeri, M.P. Singh, Toward automating crowd RE, in *2017 IEEE 25th International Requirements Engineering Conference (RE)*, (2017), pp. 512–515. https://doi.org/10.1109/RE.2017.74
3. E.C. Groen, Crowd out the competition, in *2015 IEEE first International Workshop on Crowd-Based Requirements Engineering (CrowdRE)*, (2015), pp. 13–18. https://doi.org/10.1109/CrowdRE.2015.7367583
4. E.C. Groen, S. Adam, J. Doerr, Towards crowd-based requirements engineering: a research preview, in *Requirements Engineering: Foundation for Software Quality*, (Springer, Cham, 2015), pp. 247–253
5. M. Hosseini, K. Phalp, J. Taylor, R. Ali, Towards crowdsourcing for requirements engineering, in *20th International Working Conference on Requirements Engineering: Foundations for Software Quality Empirical Track*, (2014)
6. E.C. Groen et al., The crowd in requirements engineering: the landscape and challenges. IEEE Softw. **34**(2), 44–52 (2017). https://doi.org/10.1109/MS.2017.33
7. V. Gupta, Comment on "a social network based process to minimize in-group biasedness during requirement engineering". IEEE Access **9**, 61752–61755 (2021). https://doi.org/10.1109/ACCESS.2021.3073379
8. M. van Vliet, E.C. Groen, F. Dalpiaz, S. Brinkkemper, Identifying and classifying user requirements in online feedback via crowdsourcing, in *Requirements Engineering: Foundation for Software Quality. REFSQ 2020. Lecture Notes in Computer Science*, ed. by N. Madhavji, L. Pasquale, A. Ferrari, S. Gnesi, vol. 12045, (Springer, Cham, 2020). https://doi.org/10.1007/978-3-030-44429-7_11
9. W. Maalej, M. Nayebi, T. Johann, G. Ruhe, Toward data-driven requirements engineering. IEEE Softw. **33**(1), 48–54 (2016)

Chapter 2
Towards Gamification and Crowdsourcing in Social Sector Requirements Engineering

Abstract Gamification and crowdsourcing had been widely employed in software engineering activities (especially Requirements Engineering (RE)) to leverage the power of diverse software related knowledge distributed among a crowd of motivated stakeholders. These technologies require technological platforms (for instance, social networks) to actively involve crowds for undertaking outsourced software engineering tasks. The social sector software development must ensure that software meets diverse citizen needs (participatory RE), diversity which is too diverse. To identify state of the art of crowdsourcing and gamification in social sector requirements engineering (SSRE), systematic literature survey of secondary studies focused on SE context in general and RE in particular is conducted. The holistic findings are then refined by comparing it with the limited primary studies focused on social sector context. The findings help to set the generic context and social sector context at differentiated positions. The results also provide initial indications that the social sector focused research is still in its infancy stage, and it requires considerable effort to provide crowdsourcing and gamification infrastructure to manage the user diversity in the social sector context.

Keywords Software engineering · Requirements engineering · Social sector · Crowdsourcing · Gamification

2.1 Introduction

VAMS—the Vaccine Administration Management System is developed by the consulting firm Deloitte for the US Centers for Disease Control and Prevention's. Recently it was reported that this website is giving a daunting experience to few citizens when they try to book their appointment for a coronavirus vaccine.[1] What went

[1] https://www.technologyreview.com/2021/01/30/1017086/cdc-44-million-vaccine-data-vams-problems/

V. Gupta, *Requirements Engineering for Social Sector Software Applications*, https://doi.org/10.1007/978-3-030-83549-1_2

wrong-scalability issues or quality issues? The reasons are not reported yet but if the system would have been Co-produced with diverse citizens, the results would have been different.

Take another example of Sarasota and Manatee County. Recently it was reported that the old people are feeling it really hard to take appointment to get coronavirus vaccine because appointments could only be made online.[2] What does this signify and how is it related to social sector requirements engineering (requirements engineering for designing social sector applications). This example signifies the need to provide suitable mechanisms to the user groups if they are to be involved in requirements engineering. For instance, old people cannot be given a web-based system or social platforms for taking part in crowd based requirements engineering. Requirements engineering activities need involvement of diversely large user groups to produce high quality software, but this also need customisation of activities as per user group competencies.

Social sector (also called third sector) as briefly introduced in Chap. 1 is that part of the economy that includes the activities for social good, i.e., providing benefits to the society by addressing the social problems like poverty, health issues, lack of education, hygiene, hunger, etc. The diversity of the user segments in the social sector is the biggest challenge for the RE team [1].

Requirements engineering (RE) is the software engineering (SE) sub-activity that tries to identify the real needs of the customers which are then documented as requirements after a careful analysis, prioritisation and validation. RE requires user participation to have product success in the market. User participation could be achieved either through face-to-face interactions with the RE team (called traditional RE) or through crowdsourcing platforms (called Crowd based RE). Crowdsourcing platforms help to reach a large number of users, which are globally distributed and hence not approachable through face-to-face interactions at the same physical space. Their motivation levels are boosted by the use of gamification (i.e. the use of game design elements in non-gaming context [2]).

The users of social sector applications are too diverse (for instance in their ability to access technological platforms) that existing crowdsourcing and gamification mechanisms may not be a "one size fit all" solution.

The objective of this study is to systematically review the secondary studies about gamification and/or crowdsourcing in the software engineering domain and RE in particular. The holistic findings are then refined by comparing it with the limited primary studies focused on social sector context. The findings help to set the generic context and social sector context at differentiated positions. The outcome will motivate researchers to design new solutions considering the social sector constraints, among which, diversity is the biggest challenge.

---

[2] https://eu.heraldtribune.com/story/news/local/sarasota/2021/01/04/seniors-withseniors-without-computers-may-lack-access-out-computers-may-lack-access-covid-19-vaccine/4098005001/

2.2 Tertiary Review Protocol

The tertiary study executed the systematic review guidelines as disseminated in [3] to plan review protocol, execute the protocol and report the results. This review tries to answer the following research question RQ: How is the crowdsourcing and gamification applied in software engineering in general and RE in particular?

To conduct the tertiary study, the review protocol involves triggering four bibliographic databases—IEEE Xplore, ACM digital library, SpringerLink and ScienceDirect against the search string *((“crowdsourcing”) OR (“gamification”)) AND (“software engineering”)*. The triggering resulted in 1122 results, which after being subjected to selection criteria (Inclusion and Exclusion criteria) were reduced to 05 for final synthesis.

Inclusion Criteria

- Systematically conducted literature reviews only.
- Reviews focusing on crowdsourcing and gamifications.
- Reviews focusing on generic SE and/or RE in particular.
- Studies published in 2018–2021 (both ends inclusive).
- Studies written in English and belonging to the Software Engineering domain only.

Exclusion Criteria

- Papers not able to answer either of the research question.
- Studies focusing on individual SE activity (except RE).

The execution of the tertiary review protocol resulted in 05 review studies to be subjected for further analysis to meet the research objectives (Table 2.1).

These articles were subjected to quality assessment procedures by evaluating each review studies against four formulated questions as proposed in [4]. The questions (Qi) along with the basis for rating scale of 1 (for Yes), 0.5 (for Partly) and 0 (for No) are given below.

- Are the review’s inclusion and exclusion criteria described and appropriate **(Q1)**? Y(*criteria are explicitly defined*), P(*criteria are implicitly defined*) and N (*criteria are not* defined).
- Is the literature search likely to have covered all relevant studies **(Q2)**? Y(*searching four or more bibliographic databases*), P(*searching three or four bibliographic databases*) and N (*searching up to two bibliographic databases*).

Table 2.1 Selection of studies

Bibliographic database	# Initial studies	Studies (after applying inclusion & exclusion)
IEEEXplore	261	01
SpringerLink	240	01
ACM digital library	352	00
ScienceDirect	250	03

- Did the reviewers assess the quality/validity of the included studies (**Q3**)?
  Y(*quality assessment is performed by the authors*), P(*quality issues as formulated in research question are addressed by the study*) and N (*No quality assessment is performed by the authors*).
- Were the basic data/studies adequately described (**Q4**)?
  Y(*details about individual studies are specified and it is possible to trace the summaries back to individual studies*), P(*individual papers are summarised, but it is not possible to trace back to individual studies*) and N (*results of the individual studies are not specified*).

The results of the quality assessment procedure are given in Table 2.2.

The secondary studies (05 studies) are subjected to forward snowballing by subjecting their Google citations to the Tertiary review protocol (Sect. 2.2). This resulted in 08 studies being included for further analysis (Tables 2.3 and 2.4).

Table 2.2 Quality assessment

		Quality assess.				
		Quality assessment question Y(1), P(0.5), N(0)				Total score
Review studies	Refs.	Q1	Q2	Q3	Q4	(4)
Maurício, R.D.A., Veado, L., Moreira, R.T., Figueiredo, E. and Costa, H., 2018. A systematic mapping study on game-related methods for software engineering education. Information and software technology, 95, pp.201–218	[5]	1	1	0.5	1	3.5
Alhammad, M.M. and Moreno, A.M., 2018. Gamification in software engineering education: A systematic mapping. Journal of Systems and Software, 141, pp.131–150	[6]	1	1	1	1	4
Cursino, R., Ferreira, D., Lencastre, M., Fagundes, R. and Pimentel, J. Gamification in requirements engineering: a systematic review. 2018 11th International Conference on the Quality of Information and Communications Technology (QUATIC), 4–7 Sept. 2018, pp. 119–125, Coimbra, Portugal	[7]	1	1	1	1	4
Khan, J.A., Liu, L., Wen, L. and Ali, R. Crowd intelligence in requirements engineering: Current status and future directions. International working conference on requirements engineering: Foundation for software quality, 18–21 March 2019, pp. 245–261, Essen, Germany	[8]	1	1	0.5	1	3.5
Sarı, A., Tosun, A. and Alptekin, G.I., 2019. A systematic literature review on crowdsourcing in software engineering. Journal of Systems and Software, 153, pp.200–219	[9]	1	1	1	1	4

Table 2.3 Forward snowballing and analysis

Review studies	Refs.	Citations	Number of studies selected
Maurício, R.D.A., Veado, L., Moreira, R.T., Figueiredo, E. and Costa, H., 2018. A systematic mapping study on game-related methods for software engineering education. Information and software technology, 95, pp.201–218	[5]	61	02
Alhammad, M.M. and Moreno, A.M., 2018. Gamification in software engineering education: A systematic mapping. Journal of Systems and Software, 141, pp.131–150	[6]	100	03
Cursino, R., Ferreira, D., Lencastre, M., Fagundes, R. and Pimentel, J. Gamification in requirements engineering: a systematic review. 2018 11th International Conference on the Quality of Information and Communications Technology (QUATIC), 4–7 Sept. 2018, pp. 119–125, Coimbra, Portugal	[7]	10	01
Khan, J.A., Liu, L., Wen, L. and Ali, R. Crowd intelligence in requirements engineering: Current status and future directions. International working conference on requirements engineering: Foundation for software quality, 18–21 March, 2019, pp. 245–261, Essen, Germany	[8]	14	01
Sarı, A., Tosun, A. and Alptekin, G.I., 2019. A systematic literature review on crowdsourcing in software engineering. Journal of Systems and Software, 153, pp.200–219	[9]	18	00

Table 2.3 highlights the forward snowballing details about the five secondary studies. Table 2.4 provides the details about the secondary studies selected after execution of tertiary review protocol on Google scholar citations.

These 07 studies are subjected for the quality assessment procedure with quality scores mentioned in Table 2.5. The studies scoring 02 or more points (out of 04) were only included to further analysis, leading to the further analysis of 04 studies (out of 07).

2.3 Result Analysis

2.3.1 Overview of Review Studies

Table 2.2 highlights the details about the selected ten studies (i.e., 05 parent studies and 04 child studies selected after forward snowballing). The details include publication venues hosting these review studies are well established venues for publishing high quality articles.

The details of publishing venue, number of review studies published in each venue and the year of publication are given in Table 2.6.

Table 2.4 Studies selected through forward snowballing

Studies selected	Refs.	Parent references
García-Mireles, G. A., & Morales-Trujillo, M. E. (2019, October). Gamification in Software Engineering: a tertiary study. In International Conference on Software Process Improvement (pp. 116–128). Springer, Cham	[10]	[5]
Dos Santos, A. L., Souza, M. R., Dayrell, M., & Figueiredo, E. (2018, March). A systematic mapping study on game elements and serious games for learning programming. In International Conference on Computer Supported Education (pp. 328–356). Springer, Cham	[11]	
Venter, M. (2020, April). Gamification in STEM programming courses: State of the art. In 2020 IEEE Global Engineering Education Conference (EDUCON) (pp. 859–866). IEEE	[12]	[6]
Castro, D., Arantes, F., & Werner, C. A tertiary mapping on the use of games for teaching software engineering. SBC – Proceedings of SBGames 2019. Pp: 1120–1123	[13]	
Maiga, J., & Emanuel, A. W. R. (2019). Gamification for Teaching and Learning Java Programming for Beginner Students-A Review. JCP, 14(9), 590–595	[14]	
Milosz, M., & Milosz, E. (2020, April). Gamification in Engineering Education–a Preliminary Literature Review. In 2020 IEEE Global Engineering Education Conference (EDUCON) (pp. 1975–1979). IEEE	[15]	[7]
Vogel, P., & Grotherr, C. (2020). Collaborating with the Crowd for Software Requirements Engineering: A Literature Review. MCIS 2020 Proceedings. 1. https://aisel.aisnet.org/amcis2020/virtual_communities/virtual_communities/1	[16]	[8]

2.3.2 *Results Specific to Research Objectives*

The studies were subjected to the analysis to answer the formulated research questions. The insights brought after analysis of the studies is briefly mentioned below:

2.3.2.1 How Is the Crowdsourcing and Gamification Applied in Software Engineering in General and RE in Particular?

The study [11] reported the results of the systematic mapping study of the application of serious games in learning programming. The applicability of the gamification in engineering education had been reported in [5, 6] which target the application of the gamification for the SE education of the students, which could be helpful to build skills to support crowdsourcing and gamification infrastructure for conducting RE activities. One major outcome is that the gamification had been less applied for teaching courses related to RE activities.

Authors in [7] study the gamification in the RE context highlighting the game elements used, RE activities targeted, benefits in using gamifications and associated

Table 2.5 Quality assessment

Review studies	Refs.	Quality Assess.				Total score (4)
		Quality assessment question Y(1), P(0.5), N(0)				
		Q1	Q2	Q3	Q4	
García-Mireles, G. A., & Morales-Trujillo, M. E. (2019, October). Gamification in Software Engineering: a tertiary study. In International Conference on Software Process Improvement (pp. 116–128). Springer, Cham	[10]	1	1	1	1	4
Dos Santos, A. L., Souza, M. R., Dayrell, M., & Figueiredo, E. (2018, March). A systematic mapping study on game elements and serious games for learning programming. In International Conference on Computer Supported Education (pp. 328-356). Springer, Cham	[11]	1	1	1	1	4
Venter, M. (2020, April). Gamification in STEM programming courses: State of the art. In 2020 IEEE Global Engineering Education Conference (EDUCON) (pp. 859–866). IEEE	[12]	1	1	0	1	3
Castro, D., Arantes, F., & Werner, C. A tertiary mapping on the use of games for teaching software engineering. SBC – Proceedings of SBGames 2019. pp: 1120–1123	[13]	1	0	0	0	1
Maiga, J., & Emanuel, A. W. R. (2019). Gamification for Teaching and Learning Java Programming for Beginner Students-A Review. JCP, 14(9), 590–595	[14]	0	0	0	0	0
Milosz, M., & Milosz, E. (2020, April). Gamification in Engineering Education–a Preliminary Literature Review. In 2020 IEEE Global Engineering Education Conference (EDUCON) (pp. 1975–1979). IEEE	[15]	1	0	0	0	1
Vogel, P., & Grotherr, C. (2020). Collaborating with the Crowd for Software Requirements Engineering: A Literature Review. MCIS 2020 Proceedings. 1. https://aisel.aisnet.org/amcis2020/virtual_communities/virtual_communities/1	[16]	1	1	1	1	4

Table 2.6 Publishing venues

Publishing venue	Published studies	Year	Type	Impact factor
Journal of Systems and Software	[6, 9]	2018, 2019	Journal	2.450
Information & Software Technology	[5]	2018	Journal	2.726
International Conference on Computer Supported Education	[11]	2018	Conference	
International Conference on the Quality of Information and Communications Technology (QUATIC)	[7]	2018	Conference	–
International Working Conference on Requirements Engineering: Foundation for Software Quality	[8]	2019	Conference	–
International Conference on Software Process Improvement.	[10]	2019	Conference	–
2020 IEEE Global Engineering Education Conference (EDUCON)	[12]	2020	Conference	–
Americas' Conference on Information Systems (AMCIS)	[16]	2020	Conference	–

challenges. The results indicate that different game elements are used to implement gamification (with points and leader boards being the maximum) and elicitation being targeted the most with least focus on specification activity. The gamification helps to enhance the participation thereby strengthening the RE activity (which is most communication and collaboration intensive activity) but implementation of the gamification may reduce the quality of the RE outcomes.

Author in [8] highlighted the use of crowdsourcing in RE activities by focusing on identification of RE activities focused by researchers and the various ecosystem elements that help to integrate it with RE processes. The implementation of crowdsourcing requires careful consideration about crowd, task to be gamified, mechanism, incentives to be offered and quality of the outcomes, however, in literature gamification has been applied for different RE activities. The activities include elicitation, modelling and specification, analysis and validation, prioritisation and run time monitoring.

Author in [9] presented a comprehensive overview of crowdsourcing in SE. The result indicates that coding, testing, design and requirements engineering are one of the activities focussed by the researchers. The number of studies in the literature are limited, which limits the maturity of the literature. Also, the literature lacks the studies to capture the evolution of the software.

The authors in [10] conducted a systematic mapping study to identify the state of research in the domain of gamification in software engineering. The main findings were that there is a strong need for more empirical research in gamification based software engineering. Author in [12] conducted literature review to identify the state of the art of gamification in teaching programming courses to higher education students.

Table 2.7 Summary of studies

Paper references	Research focus
[11]	Gamification in Programming
[5]	Gamification in Software Engineering Education
[6]	
[7]	Gamification in Requirements Engineering
[8]	Crowdsourcing in Requirements Engineering
[9]	Crowdsourcing in Software Engineering
[10]	Gamification in Software Engineering
[12]	Gamification in Teaching Programming
[16]	Collaborative Crowdsourcing in Requirements Engineering

Author in [16] conducted literature review to identify the state of the art of collaborative crowd based RE, i.e., application of collaborative crowdsourcing to various Requirements Engineering activities.

The findings can be summarised in Table 2.7, which represents the research focus of each study.

2.4 State of Affairs in Social Sector Context

Table 2.7 information can be arranged into 2 × 3 matrix arranged across two variables-Technology (Crowdsourcing or Gamification) and Focus (Requirements Engineering or Software Engineering or Programming). The matrix is shown in Table 2.8.

Table 2.8 highlights that the review studies review the state of the art of either gamification or crowdsourcing focusing on either areas Requirements Engineering, Software Engineering or Education (including programming).

The literature lacks the secondary studies that are focused on social sector domains. None of the 09 studies reported about applicability of crowdsourcing and gamifications in the social sector context. The lack of surveys focusing on crowdsourcing and gamifications in social sector context could be because this area is still in its infancy state and is yet to attract the researchers focus.

Limited primary studies are available that studies the Requirements Engineering in the context of the social sector. The browser-based social software platform for Social Requirements Engineering (SRE) called Requirements Bazaar was proposed in [17]. The tool allows the community members (i.e., stakeholders) to collaborate towards idea generation, idea selection and realisation. The communities can include anyone-users, designers, coders and much more. However, the secularities of the diversity in social sector users, especially less technology savvy users are not well considered.

The authors in [18] provided a vision for the mass participation of the users in requirements engineering activities using democratic concepts like delegated voting

Table 2.8 Classification of studies

Focus Technology	Requirements engineering	Software engineering	Education/programming
Crowdsourcing	[8, 16]	[9]	–
Gamification	[7]	[10]	[5, 6, 11, 12]

and structured refinement. The concepts could work well in the social sector, but the issues pertain to its implementation and integration with the different technology supported platforms. For instance, how older people can participate and how their perspectives could be merged with those who can easily access online platforms, remain uncertain.

Authors in [19] provided a vision about interrelation between requirements engineering and society. For instance, it is important to analyse the unique needs of the elder people using social media, which will felicitate the technology design decisions aligned with their needs. Requirements engineering will help to design systems that have social good. This work, however, does not provide any techniques to involve the diverse users to support social sector software development.

Author in [20] applied motivational modelling to develop a live system called Ask Izzy, a system that assists homeless Australians to identify information about the services that they need. The team was successful in understanding the needs of the special user group-those facing homelessness and conduct requirements engineering activities by overcoming the socio-technical gap. The outcome has good lessons for the requirements engineers that are involved in designing social sector solutions by involving diverse users. The issue remains with the effortless way of involving diverse users in requirements engineering, especially when they are geographically scattered and separated by diversity in knowledge, skill and competencies.

Table 2.9 gives the focus and limitations of the studies with respect to their applicability for social sector requirements engineering, i.e., requirements engineering for the software applications that are intended to be used by diverse citizens of the country.

2.5 Discussion

Author in [1] through the case study reported that diversity in the user segments of the social sector is the biggest threat to RE activity. The blended RE techniques, i.e., intermix of traditional and crowd based techniques, should be executed to identify the diverse perspectives of the user segments under limited resources (especially limited financial resources). The number of studies focusing on RE using crowdsourcing and gamifications are limited that reflects the infancy stage of the literature, which limits its ability to provide sophisticated RE techniques. Limited work in the RE area implies that at present, the social sector companies have limited

Table 2.9 Details of the social sector related studies

Study references	Focus	Limitation
[17]	Collaborative mass user participation in requirements engineering	Key issues remain unaddressed: 1. How the diverse users (especially less technology savvy users) can be involved in requirements engineering? 2. How the perspectives of different groups will be merged (if different mechanisms are used to reach different user groups).
[18]		
[19]	Requirements engineering and society	
[20]	Motivational modelling to reduce socio-technical gap	

support to get from the research disseminated in the literature. Authors in [6] reported that grounds for selecting game elements are not mentioned in the research investigated by them. Authors in [7] also highlighted that different game elements are used across different studies. Authors in [8] reported that the game design elements should match the person profile in order to bring successful motivation and participation. As the user segments are too diverse in terms of their expertise, geographical locations (beyond organisational reach) and perspectives, identifying the game elements is very effortful task. The task is further complicated due to limited support from the literature related to mapping game design with the individual user segments profiles. This means that gamification infrastructure could be a biggest validity threat affecting the quality of the RE activity outcomes.

Limited applicability of the gamification in SE education (in particular RE activity) as identified in [5, 6], limits the knowledge transfer to the SE students to gain expertise and skills specific to the social sector challenges. For instance, elicitation activity will require more practical exposure by face-to-face interactions with the user segments to identify the diversity in their perspectives, expressions, languages and expertise of using technological products. The understanding of these factors will help them to design the gamified crowdsourcing platforms that are suitable for use by diverse users and will be computationally advanced to capture diversity into single unified outcomes, i.e., validated and detailed requirements about the system. In general, different RE activities require multiple iterative interactions with the diverse and globally located users. Thus, problem domain understanding is required to provide a high level of abstraction to the users about computationally advanced systems.

Although primary studies that focus on social sector requirements engineering provide a good basis for knowledge building, yet they are too limited that could help to formulate a unified viewpoint. These studies are silent on how diverse users (including elder people) could be involved in requirements engineering decision making activities and how their perspectives will be accounted for. There is now an agreement that crowd supported requirements engineering is the focus of researchers these days but the way for ensuring the equal representation of user groups is yet to be investigated.

2.6 Conclusion and Future Work

The research in integrating crowdsourcing and gamifications for social sector RE activities is still in its infancy stage. The RE research community needs to accurately understand the social sector challenges that helps them to design the customised solution to automate RE activities. The customised solution (for each customer segment) should foster their motivation levels to participate in RE activities and simultaneously enhance the capabilities of the system to capture the user segment diversity resulting in the high quality ranked software requirements. Crowdsourcing could help to get diversely large perspectives about the software solution by motivating users through gamifications but the diversity in social sector context could limit their applicability. In the future, it is expected that the research community could integrate the social sector challenges in the formulation of crowd based RE solutions with the good game elements in it.

The study identified two research questions for the future research in crowdsourcing and gamification based social sector RE. This includes:

RQ1. How the diverse users (especially less technology savvy users) can be involved in requirements engineering?

RQ2. How the perspectives of different groups will be merged (if different mechanisms are used to reach different user groups).

References

1. V. Gupta, *Requirement Engineering Challenges for Social Sector Software Development: Insights from a Case Study*. (Communicated).
2. S. Deterding, D. Dixon, R. Khaled, L. Nacke, From game design elements to gamefulness: defining "gamification", in *Proceedings of the 15th International Academic MindTrek Conference: Envisioning Future Media Environments*, (2011), pp. 9–15
3. B. Kitchenham, S. Charters, Guidelines for performing systematic literature reviews in software engineering, Technical Report (2007)
4. B. Kitchenham, R. Pretorius, D. Budgen, O.P. Brereton, M. Turner, M. Niazi, S. Linkman, Systematic literature reviews in software engineering–a tertiary study. Inf. Softw. Technol. **52**(8), 792–805 (2010)
5. R.D.A. Maurício, L. Veado, R.T. Moreira, E. Figueiredo, H. Costa, A systematic mapping study on game-related methods for software engineering education. Inf. Softw. Technol. **95**, 201–218 (2018)
6. M.M. Alhammad, A.M. Moreno, Gamification in software engineering education: a systematic mapping. J. Syst. Softw. **141**, 131–150 (2018)
7. R. Cursino, D. Ferreira, M. Lencastre, R. Fagundes, J. Pimentel, Gamification in requirements engineering: a systematic review, in *2018 11th International Conference on the Quality of Information and Communications Technology (QUATIC)*, Coimbra, Portugal, 4–7 Sept 2018, pp. 119–125
8. J.A. Khan, L. Liu, L. Wen, R. Ali, Crowd intelligence in requirements engineering: Current status and future directions, in *International Working Conference on Requirements Engineering: Foundation for Software Quality*, Essen, Germany, 18–21 March 2019, pp. 245–261

9. A. Sarı, A. Tosun, G.I. Alptekin, A systematic literature review on crowdsourcing in software engineering. J. Syst. Softw. **153**, 200–219 (2019)
10. G.A. García-Mireles, M.E. Morales-Trujillo, Gamification in software engineering: a tertiary study, in *International Conference on Software Process Improvement*, (Springer, Cham, 2019), pp. 116–128
11. A.L. Dos Santos, M.R. Souza, M. Dayrell, E. Figueiredo, A systematic mapping study on game elements and serious games for learning programming, in *International Conference on Computer Supported Education*, (Springer, Cham, 2018), pp. 328–356
12. M. Venter, Gamification in STEM programming courses: state of the art, in *2020 IEEE Global Engineering Education Conference (EDUCON)*, (IEEE, 2020), pp. 859–866
13. D. Castro, F. Arantes, C. Werner, A tertiary mapping on the use of games for teaching software engineering, in *SBC – Proceedings of SBGames*, (2019), pp. 1120–1123
14. J. Maiga, A.W.R. Emanuel, Gamification for teaching and learning Java Programming for beginner students-a review. JCP **14**(9), 590–595 (2019)
15. M. Milosz, E. Milosz, Gamification in engineering education–a preliminary literature review, in *2020 IEEE Global Engineering Education Conference (EDUCON)*, (IEEE, 2020), pp. 1975–1979
16. P. Vogel, C. Grotherr, Collaborating with the crowd for software requirements engineering: a literature review, in *MCIS 2020 Proceedings* (2020), https://aisel.aisnet.org/amcis2020/virtual_communities/virtual_communities/1
17. D. Renzel et al., Requirements Bazaar: social requirements engineering for community-driven innovation, in *Proceedings of of IEEE RE*, (2013), pp. 326–317
18. T. Johann et al., Democratic mass participation of users in requirements engineering? in *Proceedings of IEEE RE*, (2015), pp. 256–261
19. G. Ruhe, et al., The vision: requirements engineering in society, Proceedings of IEEE RE 2017 pp. 478–479.
20. R. Burrows et al., Motivational modelling in software for homelessness: lessons from an industrial study, in *Proceedings of RE*, (2019), pp. 298–307

Chapter 3
Requirements Engineering Challenges for Social Sector Software Development: Insights from Multiple Case Studies

Abstract Requirements Engineering (RE) is a co-creation process that requires continuous involvement of the users. Diversity in the user's segments helps to provide holistic view of the expected solution from multiple perspectives to the software engineers but this diversity is a great challenge for software engineers in social sector context. The objective of this paper is to contribute to the identification of the RE processes and associated challenges in releasing the software in the social sector markets. To meet this objective, an exploratory case study is conducted with two software companies to identify the RE processes and the challenges in executing such processes with the diverse user segments. The outcome of the case study indicates that the diversity limits the ability to involve the representative samples of user populations using the same set of RE tools and techniques as one size fits all solution for all segments. The diverse user base must be partitioned into different segments, with each segment triggered using a suitable set of RE techniques, i.e. traditional and crowd based RE. The diverse perspectives learned as a result of the interaction with each segment, must be merged together into a single perspective about the software meant to be used in the social sector. There is a need for new RE process specially designed for handling the complexities of social sector, which this paper terms as Social Sector Requirements Engineering (SSRE).

Keywords Social sector requirements engineering · Social sector · Requirements engineering

3.1 Introduction

Mahiti Infotech Pvt. Ltd., a firm based in India with a focus on technological social innovation, implemented a project called "OurCrop" in the year 2012. OurCrop is a Free and Open-Source Software for the agricultural institutes to manage farming

---

This chapter appeared in ''Varun Gupta (2021) Requirement Engineering Challenges for Social Sector Software Development: Insights from Multiple Case Studies. Digit. Gov.: Res. Pract. https://doi.org/10.1145/3479982. © 2021 Copyright held by the owner/author(s)".

V. Gupta, *Requirements Engineering for Social Sector Software Applications*, https://doi.org/10.1007/978-3-030-83549-1_3

activities. This social innovation project imposed several challenges to the firm, major ones include—accurately exploring the problem domain, translating understandings into software requirements and ensuring the continuous active user engagement throughout the software development process [1]. These difficulties arose because the main stakeholders of the system were the marginalised segments of the society (i.e., farmers), which were hard to be involved for knowledge transfer with the Requirements Engineering (RE) team. There are many other social sector applications which are used by marginalised as well as technology savvy citizens (for instance, *Seguranca Social Direta* in Portugal), the diversity which makes RE much harder to execute.

RE is the software engineering sub-activity that tries to identify the real needs of the customers which are then documented as requirements after a careful analysis, prioritisation and validation. The main reason for project failures in big and small companies (like start-ups) is the inability to satisfy its intended users, i.e., delivering the features that are not required by the users [2–4]. This means that the requirement team must interact with the users to identify their needs and validate the set of requirements with them before actually started implementing them into working software. User involvement is thus a necessary condition for project success [2, 5, 6].

Traditional RE includes the activities that require users to be in close contact with the requirement team at the same place and same time, with the focus on a limited number of users [6]. The crowdsourcing based RE (called Crowd based RE) [7–9] and the Gamification versions of RE (for fostering motivation levels for continuous user participation [2, 10–14]) require the users to have skills to access the crowdsourcing infrastructure.

The social sector market is made up of the citizens as the users, which are too diverse in terms of their demographics, expectations, habits, technological experience and so on. Further they are distributed in far flung areas of the country. This diversity limits the applicability of traditional RE as face-to-face participation is not feasible with diverse users. The crowd based RE and gamification based RE are also limited as it is hard to involve the citizens with no or little technological expertise on online platforms. Furthermore, as discussed in Chap. 2, the research in this area in the social sector context is still in its infancy stage. The lack of crowdsourcing based RE techniques specific to social sector context, limits the ability of the RE team to make RE activity participatory by involving diverse users, which further lowers the chances of the social software to be successful in the market. The software solution thus fails to meet society needs and fails to deliver social goods (benefits to the society).

One major aspect in reaching out to the users for exploring the problem domain is how well the requirement team understand the market of the product (in other words, possible users) and how easily they can manage to establish communications with them? To make the accessibility easier, the role of government is a prime factor in fostering the knowledge transfer between citizens and requirement team. This is because the government has complete details about the possible users of the social sector applications, personal details about these users (for instance, data available with social institutions as provided by citizens themselves) and resources to motivate them to participate in RE activities. The success of social sector applications depends on ability to meet product/market fit, which is possible only when the citizens are accessible to the requirement team. The government processes are ICT enabled which allows them to store, process and utilise the big data about the customers. The strong focus of government to provide better services to the citizens by reducing bureaucracy using digital tools (for instance, blockchain based system for land registry) is a strong motivator for investing more in developing the software applications for social sector of the nation. This allows them to show their focus on promoting transparency, auditability, accountability, effortless and fast services. However, the adaption of these social sector applications by the citizens is probabilistic as not only they need to be convinced to use such systems but also the system should be able to match their needs. Government could act as a bridge between requirement team and citizens rather acting as the customer representative.

The objective of this paper is to investigate the various challenges experienced by the social sector software developing companies and to study their RE practices. This objective is fulfilled through an exploratory case study conducted with the two Multinational Software companies located in India and the USA. Exploratory study is conducted because RE in the social sector context is an unexplored issue in the literature. The studied companies have multiple subsidiaries located globally and are involved in delivering software products for the mass markets, including social sectors. For instance, company A product portfolio has 50% social sector product contribution and company B product portfolio has 28% social sector software products.

The study aims to answer the following research questions:

RQ1. How is RE carried out in context of social sector applications?
RQ2. What are the challenges in doing RE in context of social sector applications?

This paper is structured as follows: the case study research protocol is given in Sect. 3.2 followed by a brief background of requirements engineering techniques applicable in social sector context. Section 3.4 highlights the result of the case study which is further explained through a real user case in Sect. 3.5. This led to interesting discussions in Sect. 3.5 along with setting direction for future work. The framework for future research as derived from the cases is given in Sect. 3.7, result assessment in Sect. 3.8, implications to government and software engineers in Sect. 3.9 and finally the paper is concluded in Sect. 3.10.

3.2 Case Study Protocol

The case study is executed as per the guidelines proposed in [15]. This involves five steps, i.e., case study design, data collection procedures, collecting evidence, analysis of collected data and reporting. These steps are executed considering the validity and ethical issues.

3.2.1 Research Design

The case study is Embedded Multiple case study which studies the RE practices (**units of analysis**) in the software companies (**cases**) involved in the software development for the social sector (**context**).

The data was collected in two phases namely:

(a) Direct interviews conducted using video conferencing and
(b) Observation of the documents shared by the companies.

The insights shared in the first phase are elaborated in the second phase to address doubts, elaborate abstract details and address new perspectives that emerge after analysis of the collected data. The details of the interviewed companies and their representatives are given in Sect. 3.2. The collected data was transcribed verbatim and subjected to the grounded theory for necessary analysis.

To ensure the validity of the results, the four types of threats, i.e., construct validity, internal validity, external validity and reliability, are addressed while designing and executing the case study protocol. The collected data after the end of interview of two phases and final results were shared with the case study participants to ensure that their perspectives are accurately captured thereby addressing construct validity. The case study is exploratory in nature so internal validity is not a threat. The use of multiple levels of data collections through two methods, i.e., interviews and archival analysis and providing the chain of evidence addresses reliability.

The data shared by the company representatives comes from their wide experience in developing social sector applications, which could be meaningful for other companies as well. As the companies are diverse in terms of their working context and resources so the findings could not be much interesting those marginal cases and hence external validity is marginally impacted.

3.2.2 Case Details

The cases in this case study were selected because of the following reasons:

(a) The companies had been in the market from past many years (and thus have consortium of product release experiences as their strategic assets).
(b) Have trace record of successful market software projects.
(c) Have been successful in delivering software in the social sector (as a part of their have social corporate responsibility and for product diversifications in social sector markets).

These companies have multiple subsidiaries located globally and are involved in delivering software products for the mass markets, including social sectors. For instance, company A product portfolio has 50% social sector product contribution, and the percentage is 28% for the company B.

Table 3.1 Case details

S. No.	Company name	Main location of cases	Other locations	Software products
1.	A	India	Global	Large Portfolio
2.	B	USA	Global	Large Portfolio

Table 3.1 gives the characteristics of such companies without revealing their identity. The company representatives agreed to participate on the condition of anonymity, so company names are replaced with A and B.

The case study involved two representatives from each company. The representatives were senior managers who had diverse experience in leading the social sector project commercialisation. Their participation brought diverse perspectives driven by their experiences with different projects exhibiting different roles.

Table 3.2 gives the brief information about the social sector projects undertaken by the studied companies. The number of social sector projects implemented by the companies counts each incremental version of the software delivered to the society as a single individual project.

3.3 Background

Gamification and crowdsourcing had been widely employed in software engineering activities (especially Requirements Engineering (RE)) to leverage the power of diverse software related knowledge distributed among a crowd of motivated stakeholders. These technologies require technological platforms (for instance, social networks) to actively involve crowds for undertaking outsourced software engineering tasks. In context of RE, crowdsourcing help the requirement analysts to involve the globally distributed stakeholders (especially the users) to provide their feedback (or requirements).

The requirements provided by the stakeholders are then prioritised by them through their social interactions in the form of voting, comments, etc. [16, 17]. Crowdsourcing platforms help to reach a large number of users, which are globally distributed and hence not approachable through face-to-face interactions at the same physical space. Their motivation levels are boosted by the use of gamification.

The literature provides good references to the work already reported in the domain of crowdsourcing based RE [18–29]. The overall analysis of the published literature suggest that the crowdsourcing had been applied in generic settings without being focused on specific application domain (for instance, social sector or public administration). The crowd based RE had been witnessing the challenges like ability of the crowdsourcing platform to integrate feedbacks from different feedback channels (for instance, combining user feedbacks across App stores and social networks), efficiently processing the highly unstructured and noisy user feedbacks expressed in natural languages, ability to trust the feedback provider without compromising his privacy in crowdsourcing platforms (and filtering out those expressed

Table 3.2 Project details of studied companies

S. No.	Company name	End users	Problem domain targeted	Total projects in portfolio	Number of social sector projects	% Social sector projects
1.	A	Social sector institutions	Social media analytics software	68	20	50%
			Ideas evaluation			
			Automation of activities			
			Tax management			
		Citizens	Child welfare		14	
			Medical support to poor families			
			Education			
			Woman entrepreneurship			
2.	B	Social Sector Institutions	Automation of activities	142	32	28%
			Funding request applications			
			Audit reports			
			Funding management			
			Crowdfunding campaign management			
			Scheme social impact analytics			
			Tax management			
		Citizens	Education		08	
			Homeless people management			
			Sustainability (environment)			

by illegitimate users), aggregating the feedback responses (collective intelligence) of the providers to support RE decisions. Building over these limitations, the applicability of the crowdsourcing techniques for RE in social sector application domain is still in its infancy stage.

The literature has limited studies pertaining to requirements engineering in social sector context. The limited studies do address the requirements engineering in social sector context, but they do not provide rigorous solutions that addresses the unique challenges of social software applications.

Authors in [30] proposed Requirements Bazaar—A browser-based social software platform for Social Requirements Engineering (SRE) that allow community members to have idea generation, idea selection and realisation collaboratively. The authors in [31] provided an opinion about participation of the user masses in requirements engineering activities using delegated voting and structured refinement. Authors in [32] provided an opinion about interrelation between requirements engineering and society; with requirements engineering uncovering the needs of society. For instance, it is important to analyse the unique needs of the elder people using social media, which will felicitate the technology design decisions aligned with their needs. Author in [33] applied motivational modelling to develop a live system called Ask Izzy, a system that assists homeless Australians to identify information about the services that they need. The team was successful in understanding the needs of the special user group-those facing homelessness and conduct requirements engineering activities by overcoming the socio-technical gap. The outcome has good lessons for the requirements engineers that are involved in designing social sector solutions by involving diverse users.

Table 3.3 gives the focus and limitations of the studies with respect to their applicability for social sector requirements engineering-, i.e., requirements engineering for the software applications that are intended to be used by diverse citizens of the country.

Table 3.3 highlights that there is a need to better understand the challenges faced by requirement analysts in exploring the problem domain of social sector applications. The understanding will help to future drive the research addressing the social challenges in requirements engineering.

3.4 Result Analysis

The data collected from the cases are subjected to analysis to generate answers to the formulated research questions. The RE challenges specific to the social sector application are individually mentioned under individual research questions below:

3.4.1 RQ1 How Is RE Carried Out in Context of Social Sector Applications?

The software companies usually undertake bespoke (single client like NGO or the Government) and/or mass market software development (wider market) of the social sector software. Both studied companies perform the RE activities in flexible, light weight processes, jointly with the users. The users are highly diverse in terms of their technical expertise, language, ages, geographical locations, etc. This limits the involvement of the crowd of users throughout the RE activities. Furthermore, the user

Table 3.3 Details of the social sector related studies

Study references	Focus	Research type	Suitability for social sector				Limitation
			Diversity of user	Geographical access	Computational intelligence from diverse perspectives	Non-functional aspects like scalability, ease of use, etc.	
[16]	Collaborative mass user participation in requirements Engineering	New working method	No	Yes	No	No	Key issues remain unaddressed: How the diverse users (especially less technology savvy users) can be involved in requirements engineering? How the perspectives of different groups will be merged (if different mechanisms are used to reach different user groups)
[17]		Vision	No	No	No	No	
[32]	Requirements Engineering and Society	Vision	No	No	No	No	
[33]	Motivational modelling to reduce socio-technical gap.	New working method	No	No	No	No	

segments are beyond the reach of the organisation as they could be globally dispersed. For example, one of the respondents from company A, "*Social sector application may be intended to be offering services to the senior citizens. Their health issues, large ages, technical expertise, etc. limits their continuous involvement throughout RE. However, the best we could do is to interact to the maximum with them to convert the interaction into the knowledge that drives the engineering process*".

The users specify the changes or make new requests by specifying the problems they are facing (i.e., the pain points). The RE team prefer to observe and interact with the users to enhance the understanding of the problem domain. This understanding helps them to map the problems into software requirements, which are validated in an informal discussion with the users. As per one of the respondents from company B, "*Due to diversity in the users and lack of financial resources for social projects, we need to strike a balance between value to be offered and the efforts to be invested for such offerings. That is the reason why we execute RE activities as light weight processes driven by accurate & validated understanding of the customer problem domain*". understanding about the problem domain and the users.

The manner the RE activities are conducted by the companies are mentioned in Table 3.4.

The level of involvement of the users for different RE activities varies as shown in Fig. 3.1 (number of users employed vs. individual RE activities). The graphical representation is drawn based on the average project data of the two companies (14 projects implemented by the company A and 08 projects implemented by company B).

Table 3.4 RE activities

S. No.	RE activity	Process involved
1.	Requirement gathering	Informal discussions and observations
2.	Requirement analysis	Manually by requirement analysts
3.	Requirement prioritisation	By team of requirement analysts and product manager on the basis of their gut feeling (understanding of the problem domain)
4.	Requirement validation	Through prototypes (animations) which are validated with the sample of users
5.	Requirement documentation	Done informally using computer notes or sticky notes on the wall
6.	Requirement evolution	Discussion and observation of the set of users who requested for the changes in the requirements (by expressing their problems)

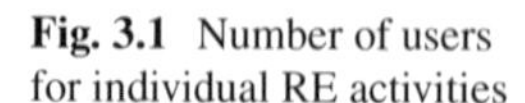
Fig. 3.1 Number of users for individual RE activities

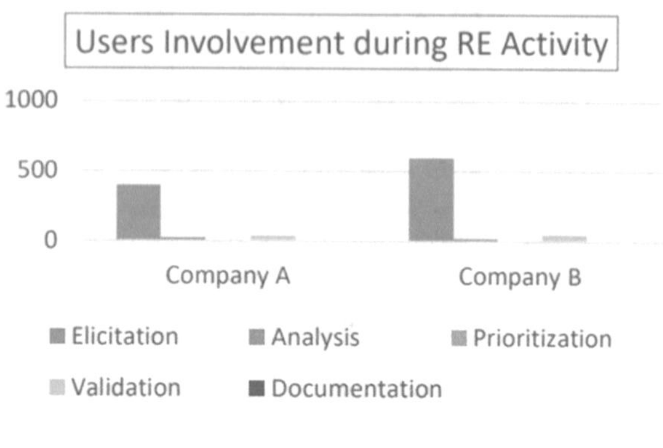

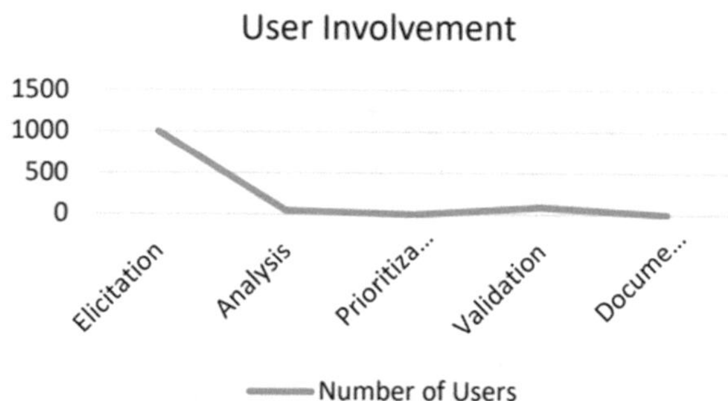

Fig. 3.2 Variation of User Involvement throughout RE for both companies

As shown, Fig. 3.2 is another equivalent representation of Fig. 3.1, which shows graphically the variation of the user involvement for each RE activity, for the studied companies only. In Fig. 3.2, Y axis denote the number of users employed for different RE phases, represented by X axis.

3.4.2 *RQ2 What Are the Challenges in Doing RE in Context of Social Sector Applications?*

RE is the co-creation process that must involve users throughout the whole process. However, the RE team must cope up with numerous issues while executing the RE activity as mentioned below:

(a) Complex and unexplored problem domain**:** The problem domain is unclear initially. This is because the end users (citizens) do not know exactly their needs. Finding an access to good representatives of these end users (because many may not be known initially) and making them agree to participate in RE, is a greater challenge for the requirement team. Understanding the social problems is harder exercise, the expertise which companies lack otherwise (for instance, social science researchers). The support of secondary data is minimal as they could be less accurate and may not fulfil the purpose. The companies collect user pain points from the NGOs and the volunteers, to get an understanding of the problem domain.

(b) **Diversity of Users:** User Diversity is the source of diverse perspectives about the solution, provided that establishing crowdsourcing infrastructure is possible. This is, however, a major challenge in the case of the social sector. This is because, social sector applications must deliver benefits to all eligible sections of the society irrespective of any bias. Minority segments have equal representation in the solution and the overall success of the system. The diverse segments are further harder to be involved in RE activities both in the company premises or through crowdsourcing platforms. For instance, crowdsourcing will not be feasible with less technology savvy people especially uneducated, senior citizens and so on. The diversity thus inhibits the co-creation RE process.

(c) Financial restrictions**:** The social sector applications are either purchased by single customer (bespoke) or purchased by many customers (mass market). However, their objective is to purchase software for social good, i.e., to provide benefits to the social sector (i.e., people) without any sake of the profit. Thus, the purchases are made through grants or donation funds, which are limited. This puts pressure on the software development process as the cost overrun could result in project failures.

The financial restrictions will be the greatest inhibitor for the small software companies compared to the big companies, to deliver software for the social good. The reason is that RE is a costly activity as it requires frequent interactions with the globally dispersed and diverse customers, on the continuous basis. The area of RE in social sector context is yet to gain attraction so to succeed, the companies should continuously perform experimentations with the users to get validated learning and invest heavily in Research and Development activities, which seems to be feasibly in big companies compared to the small ones.

(d) **Limited scope for** crowdsourcing **of diverse perspectives:** Due to large diversity in the users of social sector applications in terms of their age, education and experience, it is difficult to access their understanding of the problem domain through crowdsourcing platforms. Furthermore, it is hard to motivate them to continuously use crowdsourcing platforms because of their ignorance of the benefits to be offered by the system (because of the lack of understanding of the system domain).

Further if the crowdsourcing seems to be promising with few user segments (for instance, the college students), even though the feedback will be noisy and diverse in the way it is expressed, which makes their automated analysis using natural language processing techniques, quite difficult.

(e) **Role of** Non-functional requirements**:** Social sector applications are to be used by the citizens for solving their social problems. The user base is highly diverse and for the solution to be absorbed among them, it should satisfy non-functional requirements like easy to use, high performance, security, privacy, etc. The ability of the software to satisfy the users not only depends on their functional utilities but also on the non-functional aspects of the software solution, which actually determine the absorption of the innovation among citizens. Furthermore, identifying non-functional aspects is based on accurate analysis of the user segments (for instance, their familiarity with the use of smart phones), which is hard to be made in geographically located diverse user segments.

As per one of the respondents from company B, "*we were designing a social sector application for the senior citizens. Our observation during the requirement gathering stage helped us realise that 50% of them do not know how to use smart phones and could use their mobiles for simple functions only. This helped us realise that the offered solution should have less keyboard-based inputs. We offered authentication based on fingerprints and the functionality using simple touch based Graphical User Interface*".

For instance, the social sector applications access the private personal information of the citizens and are associated with the monetary benefits. This needs strong security features to avoid any breach of the system security mechanisms. Also, due to different expertise levels of the citizens and different devices they use, it is very challenging to find a solution that works optimally across all the platforms.

(f) **Technological solution** Integration Issues**:** During Requirements engineering, the efforts are being made to understand the software requirements including non-functional ones. The major challenge is that in many cases, the software solution architecture requires integration with the other applications or databases managed by the government. This requires designing the system on the assumption of positive outcome of the bureaucratic processes involved in gaining permission to access the information from centralised databases.

(g) Software evolution**:** The social sector application must be evolved continuously because in the social sector context, the environment is highly volatile. For instance, the government policies always change. This needs the highly flexible RE process to map new perspectives into software requirements and ensuring that further evolutions are not limited by the presence of the technical debts.

3.5 Real Use Case

The software application in the problem domain of "Medical support to poor family's application" aims to connect the poor families (called beneficiary) with the single centralised system that could take care of their medical treatments. This software application is implemented by the company A.

The user base was too diverse as this included poor families, poor senior citizens (with no children), poor orphan children, etc. Furthermore, the number of users is too large quantitatively and dispersed across the nation, which limits the ability of the requirement team to approach them for requirement gathering. The bad financial situation prevailing among these users signifies the lack of ownership of the smartphones and high-speed internet facilities which limits the ideas of using crowdsourcing for requirement gathering.

The requirement team decided to contact the local NGOs to gain better understanding of the problem domain. This proved to be beneficial as diverging perspectives about the problem domain came out of the interactions. Based on the interactions, the team found access to the few samples of the user base, which were interviewed through face-to-face traditional requirement gathering techniques, with a focus on understanding their pain points.

More interactions brought more insights like difficulty in gaining access to the specialised treatments, too much bureaucratic processes in calming financial support for medical treatment, etc., which were not mentioned by the NGOs. This highlights the gaps between the perspectives of the stakeholders including actual users. As per one of the respondents from company A, "*The single customer never understands the actual needs of the intended users and In fact they strongly trust their understanding about problem domain, which are just unvalidated assumptions*".

Some of the user's segments were also interacted using online video conferencing techniques, with remote NGO taking responsibility for arranging such meetings. The support of the volunteers in observing the poor people and interacting with them, helped the requirement team to broaden their understanding about the problem domain. However, it was difficult to interact with them initially because they were suspicious about the activities of the requirement team, and they lacked motivation because they were not confident that the solution would change their lives. It was also hard to motivate the users to participate in the requirement gathering activities. As per one of the respondents of company A, "*The hardest part is to continuously involve the users in the RE activities because they do not see value in the product in the first instant which limits their motivation levels*".

The software solution for the identified problem was confronted with numerous challenges. This includes (a) the present mobile phones used by users were simple phones and did not support the mobile applications, (b) users have low expertise with the accessing phones for special tasks, i.e. making an access to the mobile apps, which means that solution should provide a simple "touch and select" type access, (c) the eligibility of the beneficiaries is to be verified from government records which require access to the central database that incurs privacy concerns, (d) security is to be provided to avoid breach of financial information or wrong debits of the funds, (e) application should support non-functional features also to promote their adoption among the users.

Finally, the requirement team decided that the software solution will register the beneficiaries validated information with the help of the local NGOs and volunteers. The software solution will be mobile apps which will be accessible to the local NGOs (except the privacy information of the beneficiary). The mobile number of the beneficiary will be uniquely linked to his financial and medical records. A simple call to the number or simple message, will issue him/her a ticket, which means that he is registering for the medical payment related requests. The request is handled by local NGO and volunteers. The system has support for crowdfunding for managing the funds for the charitable cause.

3.6 Discussion and Future Directions

The diversity and size of the users of the social sector applications is a challenge for evolving the software product. It is well known that the software must be continuously innovated in order to keep satisfying the users because of the changes in the user needs and change in the environments (for instance, government regulations, policies, etc.). The changes in the environment happen very frequently, which prove a litmus test for the evolution of the software. Mapping the diverse user perspectives in the software requirements require face-to-face interactions with the users and strong abilities to map non-technical aspects (user insights) into technical ones (technical requirements).

The emerging problem solving paradigms like crowdsourcing and gamified crowdsourcing (with game elements to boost motivations) is not a one size fit all solution for all segments of the users. These paradigms could work perfectly fine for the segments that are technology savvy or have experience of using the technological platforms. For the remaining segments, different traditional RE techniques involving face-to-face interviews, non-technical prototypes and observations is a must. This means that RE must be executed in the blended way, involving the traditional RE and the one involving emerging problem solving paradigms. The efforts need to be made to merge the perspectives brought by traditional and new RE processes, which is another litmus test.

In future, following two aspects are worth investigating. First is to investigate the new RE processes suitable for those user segments that are not accessible through the latest technologies. This provides several challenges to address, including:

(a) Efficient ways to map non-technical perspectives into technical perspectives that are suitable for software development teams to work with.
(b) Identification of suitable prototype solutions that have the ability to extract as much as possible information about the problem domain followed by the validation.
(c) Identifying the ways of involving the users to prioritise the software requirements. This includes a challenge to present requirements in a non-technical way to the users to get their preferences and then to map these preferences into ranking decisions. Ranking decisions should also resolve trade-offs with the software development team perspectives.

The second aspect is to address issues with the user segments that are accessible through the latest technologies (or techniques) like Crowd based RE, including the following:

(a) Designing gamification that motivates different users in the same segment. For example, socialisers will be more motivated if game elements provide them the ability to enhance social interactions.
(b) Designing crowdsourcing platforms with the ability to collect the user ideas, promote social interaction and convert the interactions into RE decisions like prioritised list of software requirements. The automated natural language solu-

tions for analysing the crowd inputs and social interactions will affect the accuracy of the final solutions. This is further hindered because the users will provide their feedback in different expressions, different terminologies and at different levels of details. On one hand user privacy needs to be respected and on the other, the fake user's need to be identified with their contributions filtered out.

(c) Merging the perspectives of the users with those that could not be accessed using technologically advanced RE platforms is computationally hard.

The overall objective is to provide software companies the ability to execute the accurate and trustworthy RE process model that is lightweight and flexible. The need is to address research challenges with the emerging RE solutions and with the traditional processes, to provide a blended solution that has the ability to make RE a pure co-creation process with the users.

3.7 Framework

The results of the case study bring a useful framework that could set the boundaries for the future research in Social Sector Requirements Engineering (SSRE)-Requirements Engineering for Social Sector Applications. The SSRE Framework is represented by Fig. 3.3.

The framework represents the following elements:

(a) **Processes:** This represents the RE processes which is the consortium of diverse techniques customised for each user segment. The mapping processes converts the non-technical perspectives of the users into software requirements (technical). The diverse perspectives of user segments as gathered during RE are merged together into system understanding by the integrative processes.

(b) **People:** The people involved in the SRE include diverse user segments, requirement analysts and domain experts. Involvement of domain experts helps the RE analysts to better understand the perspectives shared by some user segments

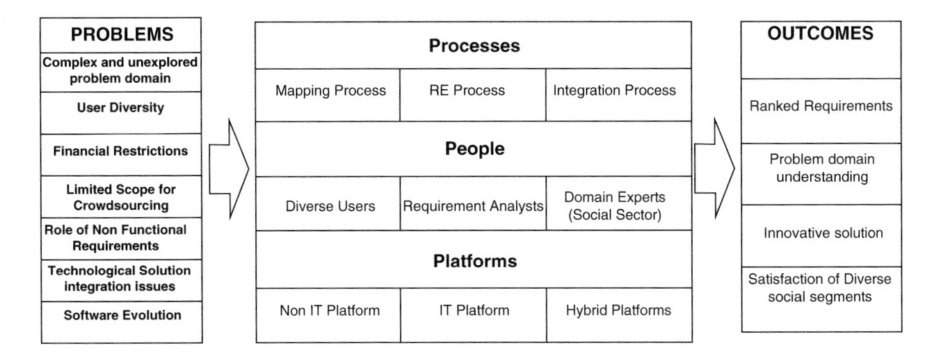

Fig. 3.3 SSRE Framework

(for instance, by farmers in their local languages) and also enhance their understanding about the social sectors.

(c) **Platforms:** To cater to the needs of the diverse users, the SRE will have Non-IT based mechanisms (for instance, face-to-face interactions), IT Platforms (For instance, online web application with crowdsourcing and gamification features) and Hybrid mechanisms (for instance, web application with online interactions for non-technology savvy users and social interactions for technology savvy users).

The overall outcome of SSRE is the better understanding of the social sector problem domain (which helps in future software evolutions) and ranked list of requirements that have the capacity to satisfy the diverse user segments.

3.8 Result Assessment

To ensure that the data collected in the case study is both valid and accurate, the member checking was performed with the company A and B. Total eight employees of both companies participated in the result assessment (more than the number of case study participants). This ensures that data is correctly collected and analysed as shared by the company representatives and the company representatives shared correct data.

The post case study questionnaire was shared with the company employees. The rating scale of 1 (not agree) to 5 (strongly agree) was used to specify the employee's ratings about the shared information. The aggregated collected responses are given in Table 3.5, which shows that a majority of the employees agreed to the validity of the results.

3.9 Implications to Government and Software Engineers

The social sector reforms had traditionally been the responsibility of the governments. Public sector institutions are held responsible for providing their services for removing social problems in order to benefit the community. Government

Table 3.5 Result assessment

Question	1	2	3	4	5	Agreed?
Diversity is the biggest threat	0	0	0	1	7	Yes
Blended requirements engineering techniques is the solution in social sector	0	0	0	0	8	Yes
All constraints are correctly mentioned	0	0	1	2	5	Yes
Requirements engineering practices are correctly shared	0	0	0	0	8	Yes
User involvement decreases from requirement gathering activity onwards	0	0	0	0	8	Yes

institutions undertake social innovation through the adoption of technology for which they outsource their needs to third party software development companies (including public undertakings). The software firms are having best expertise in undertaking technical work, but they lack the deeper understanding of the societal challenges (expect those that they observe in daily lives).

The accessibility of the community will be a challenge for the requirement analysts. High quality software will provide value to the citizens that indirectly will be valuable to the governments in meeting their objectives. The ability of the solution to meet the needs of the community and getting accepted by them will require a requirement analyst to be familiar with social problems. For this to happen, the role of government is quite important. The government support will not only help software firms to develop good quality software, but it will provide the research community with access to diverse experiences that will drive future research resulting in rigorous requirements engineering solutions.

3.10 Conclusions

The exploratory case study identified the RE process executed by the software development companies and the associated challenges in the continuous delivery of the high quality software in the social sector markets. RE must be able to capture the diverse user perspectives continuously to keep innovating the software for the social good. Maintaining continuous interactions with the social sector user segments is must for RE but their diversity is a biggest challenge for their continuous involvement for evolutionary RE. The tools and techniques used in traditional or crowd based RE will not work for all user segments in social context. In fact, a blended mix of traditional and emerging techniques should be used to converge the divergent perspectives of diversely large users into a single perspective about solution. The success of the SSRE depend on joint coordination between software engineers and the government institutions.

References

1. P. Bhatt, A.J. Ahmad, M.A. Roomi, Social innovation with open source software: user engagement and development challenges in India. Technovation **52**, 28–39 (2016). https://doi.org/10.1016/j.technovation.2016.01.004
2. R. Snijders, F. Dalpiaz, S. Brinkkemper, M. Hosseini, R. Ali, A. Ozum, REfine: a gamified platform for participatory requirements engineering, in *2015 IEEE First International Workshop on Crowd-Based Requirements Engineering (CrowdRE)*, Ottawa, ON (2015), pp. 1–6. https://doi.org/10.1109/CrowdRE.2015.7367581
3. N. Paternoster, C. Giardino, M. Unterkalmsteiner, T. Gorschek, P. Abrahamsson, Software development in startup companies: a systematic mapping study. Inf. Softw. Technol. **56**(10), 1200–1218 (2014). https://doi.org/10.1016/j.infsof.2014.04.014

4. M. Cantamessa, V. Gatteschi, G. Perboli, M. Rosano, Startups' roads to failure. Sustainability **10**, 2346 (2018). https://doi.org/10.3390/su10072346
5. F. Dalpiaz, R. Snijders, S. Brinkkemper, M. Hosseini, A. Shahri, R. Ali, Engaging the crowd of stakeholders in requirements engineering via gamification, in *Gamification*, (Springer, Cham, 2017), pp. 123–135. https://doi.org/10.1007/978-3-319-45557-0_9
6. A. Menkveld, S. Brinkkemper, F. Dalpiaz, User story writing in crowd requirements engineering: the case of a web application for sports tournament planning, in *2019 IEEE 27th International Requirements Engineering Conference Workshops (REW)*, Jeju Island, Korea (South) (2019), pp. 174–179. https://doi.org/10.1109/REW.2019.00037
7. E.C. Groen, N. Seyff, R. Ali, F. Dalpiaz, J. Doerr, E. Guzman, M. Hosseini, J. Marco, M. Oriol, A. Perini, M. Stade, The crowd in requirements engineering: the landscape and challenges. IEEE Softw. **34**(2), 44–52 (2017). https://doi.org/10.1109/MS.2017.33
8. J.A. Khan, L. Liu, L. Wen, R. Ali, Crowd intelligence in requirements engineering: current status and future directions, in *Requirements Engineering: Foundation for Software Quality. REFSQ 2019. Lecture Notes in Computer Science*, ed. by E. Knauss, M. Goedicke, vol. 11412, (Springer, Cham, 2019). https://doi.org/10.1007/978-3-030-15538-4_18
9. R. Sharma, A. Sureka, CRUISE: a platform for crowdsourcing requirements elicitation and evolution, in *2017 Tenth International Conference on Contemporary Computing (IC3)*, Noida (2017), pp. 1–7. https://doi.org/10.1109/IC3.2017.8284308
10. M.Z. Kolpondinos, M. Glinz, GARUSO: a gamification approach for involving stakeholders outside organizational reach in requirements engineering. Requir. Eng. **25**, 185–212 (2020). https://doi.org/10.1007/s00766-019-00314-z
11. R. Cursino, D. Ferreira, M. Lencastre, R. Fagundes, J. Pimentel, Gamification in requirements engineering: a systematic review, in *2018 11th International Conference on the Quality of Information and Communications Technology (QUATIC)*, Coimbra (2018), pp. 119–125. https://doi.org/10.1109/QUATIC.2018.00025.
12. M.Z. Huber Kolpondinos, M. Glinz, Behind points and levels — the influence of gamification algorithms on requirements prioritization, in *2017 IEEE 25th International Requirements Engineering Conference (RE)*, Lisbon (2017), pp. 332–341. https://doi.org/10.1109/RE.2017.59.
13. F.M. Kifetew, et al., Gamifying collaborative prioritization: does pointsification work?, in *2017 IEEE 25th International Requirements Engineering Conference (RE)*, Lisbon (2017), pp. 322–331. https://doi.org/10.1109/RE.2017.66.
14. F. Kifetew, D. Munante, A. Perini, A. Susi, A. Siena, P. Busetta, DMGame: a gamified collaborative requirements prioritisation tool, in *2017 IEEE 25th International Requirements Engineering Conference (RE)*, Lisbon (2017), pp. 468–469. https://doi.org/10.1109/RE.2017.46
15. P. Runeson, M. Höst, Guidelines for conducting and reporting case study research in software engineering. Emp. Softw. Eng. **14**(2), 131 (2009)
16. V. Gupta, Comment on "A social network based process to minimize in-group biasedness during requirement engineering". IEEE Access **9**, 61752–61755 (2021). https://doi.org/10.1109/ACCESS.2021.3073379
17. S. Mughal, A. Abbas, N. Ahmad, S.U. Khan, A social network based process to minimize in-group biasedness during requirement engineering. IEEE Access **6**, 66870–66885 (2018). https://doi.org/10.1109/ACCESS.2018.2879385
18. U.S. Ghanyani, M. Murad, W. Mahmood, Crowd-based requirement engineering. Int. J. Educ. Manage. Eng. **3**, 43–53 (2018)
19. E.C. Groen, J. Doerr, S. Adam, Towards crowd-based requirements engineering a research preview, in *Requirements Engineering: Foundation for Software Quality. REFSQ 2015. Lecture Notes in Computer Science*, ed. by S. Fricker, K. Schneider, vol. 9013, (Springer, Cham, 2015). https://doi.org/10.1007/978-3-319-16101-3_16
20. S. Taj, Q. Arain, I. Memon, A. Zubedi, To apply data mining for classification of crowd sourced software requirements, in *Proceedings of the 2019 8th International Conference on Software*

and Information Engineering (ICSIE '19), (Association for Computing Machinery, New York, 2019), pp. 42–46. https://doi.org/10.1145/3328833.3328837
21. A. Menkveld, S. Brinkkemper, F. Dalpiaz, User story writing in crowd requirements engineering: the case of a web application for sports tournament planning, in *2019 IEEE 27th International Requirements Engineering Conference Workshops (REW)*, (IEEE, 2019), pp. 174–179
22. D.A.P. Sari, A.Y. Putri, M. Hanggareni, A. Anjani, M.L.O. Siswondo, I.K. Raharjana, Crowdsourcing as a tool to elicit software requirements, in *AIP Conference Proceedings*, vol. 2329, No. 1, (AIP Publishing LLC, 2021), p. 050001
23. A. Adepetu, A.A. Khaja, Y. Al Abd, A. Al Zaabi, D. Svetinovic, Crowdrequire: a requirements engineering crowdsourcing platform. In *2012 AAAI Spring Symposium Series* (2012)
24. M. Hosseini, A. Shahri, K. Phalp, J. Taylor, R. Ali, F. Dalpiaz, Configuring crowdsourcing for requirements elicitation, in *2015 IEEE ninth International Conference on Research Challenges in Information Science (RCIS)*, (2015), pp. 133–138. https://doi.org/10.1109/RCIS.2015.7128873
25. R. Snijders, F. Dalpiaz, M. Hosseini, A. Shahri, R. Ali, Crowd-centric requirements engineering, in *2014 IEEE/ACM 7th International Conference on Utility and Cloud Computing*, (IEEE, 2014), pp. 614–615
26. C. Li, L. Huang, J. Ge, B. Luo, V. Ng, Automatically classifying user requests in crowdsourcing requirements engineering. J. Syst. Softw. **138**, 108–123 (2018)
27. T. Ambreen, Handling socio-technical barriers involved in crowd-based requirements elicitation, in *2019 IEEE 27th International Requirements Engineering Conference (RE)*, (2019), pp. 476–481. https://doi.org/10.1109/RE.2019.00065
28. P.K. Murukannaiah, N. Ajmeri, M.P. Singh, Toward automating crowd RE, in *2017 IEEE 25th International Requirements Engineering Conference (RE)*, (2017), pp. 512–515. https://doi.org/10.1109/RE.2017.74
29. E.C. Groen, Crowd out the competition, in *2015 IEEE 1st International Workshop on Crowd-Based Requirements Engineering (CrowdRE)*, (IEEE, 2015), pp. 13–18
30. D. Renzel, et al., Requirements Bazaar: social requirements engineering for community-driven innovation, in *Proceedings of IEEE RE* (2013), pp. 326–317
31. T. Johann, et al., Democratic mass participation of users in requirements engineering? in *Proceedings of IEEE RE* (2015), pp. 256–261.
32. G. Ruhe, et al., The vision: requirements engineering in society, in *Proceedings of IEEE RE 2017*, Sept 2017, pp. 478–479.
33. R. Burrows, et al., Motivational modelling in software for homelessness: lessons from an industrial study, in *Proceedings of RE* (2019), pp. 298–307.

Chapter 4
Social Sector Requirements Engineering Process Using Customer Journeys

Abstract The research chapter proposes a crowdsourcing and gamification based Requirements Engineering (RE) process that help software engineers to identify the features for the social sector applications and rank them. Requirements engineering activities are conducted by converging the diverse perspectives of the heterogenous crowds (i.e., citizens), which are actual users of the social sector software application. The algorithm elicits the customer journeys from the users (i.e., the citizens), which are then subjected to further elaboration and prioritisation by initiating the social interactions between the users. The proposed solution is evaluated using the real case of the social sector application which was executed for social welfare by one of the leading Indian NGO. The NGO activists were involved to provide software features along with their perspectives about their priorities which were compared with the benchmark results; the results represented by the 1-year execution history of the live social sector application.

Keywords Crowdsourcing · Gamification · Requirements engineering · Social sector requirements engineering · Social entrepreneurs

4.1 Introduction

Requirements Engineering (RE) is the software engineering sub-activity that tries to identify the real needs of the customers which are then documented as requirements after a careful analysis, prioritisation and validation. The objective of RE is document the requirements of the proposed software solution.

Social sector (also called third sector) is that part of the economy that includes the activities for social good, i.e., providing benefits to the society by addressing the social problems like poverty, health issues, lack of education, hygiene, hunger, etc. Social innovation has been the focus of government these days; innovation that is absorbed in the society through technological solutions. Technological solutions achieving social innovation need software engineers to explore the problem domain

V. Gupta, *Requirements Engineering for Social Sector Software Applications*, https://doi.org/10.1007/978-3-030-83549-1_4

to better understand the social challenges; to identify the value proposition that best addresses the social needs. The mapping between the technological solutions and actual customer needs (which happens to be government institutions and/or citizens) is a challenging activity, owing to the diversity in the actual users [1]. As discussed in Chap. 3, the user diversity is the major inhibitor in doing the participatory RE in social sector as it limits the applicability of traditional RE as face-to-face participation is not feasible with diverse users. Adding to this, the lack of crowdsourcing-based solutions, specific to social sector context, as discussed in Chap. 2, also inhibits the RE activities.

Software companies delivering software for social start-ups (or Small and Medium Enterprises, SMEs or even bigger companies) to conduct their social activities (for instance, delivering food packets to needy people using automated tracking and distribution software system) have to balance a trade-off between product/market fit and development cost (which determines selling price). Product/market fit requires participatory RE activities to be conducted with diverse users and cost optimisation depends on the amount of resources invested in the development process. Balancing these two aspects need an optimal process for RE that is scalable enough for a crowd of users. The user participation is required for the success of the software solution in the market. To facilitate the identification of software requirements from diverse user perspectives, the requirement team involves a crowd of users for requirement elicitation through online platforms (for instance, crowdsourcing platforms); the activity commonly referred to Crowd based RE [2–4]. To further improve the participation of the users, the requirement team uses game elements in requirements engineering; activity called gamification [5–10].

Authors in [1] identified that diversity in the user segments is the biggest challenge to conduct RE activities in the social sector domain; this is based on the reason that a single set of techniques, tools and processes does not work with all types of users; defining a new term called Social Sector Requirements Engineering (SSRE). For instance, social networks cannot be used with elder people, who are less technology savvy and hence cannot use technologically advanced solutions. Traditional RE that involve same physical space presential face-to-face interaction needs is not feasible with geographically distributed group of users.

The rich social interactions between the users by posting, commenting and rating the requirements results in a better requirement set as it captures the diverse crowd perspectives. However, social interactions sometimes lead to false positives because commenting, ratings and posting activities are based on one's perspectives based on understanding of others stated needs. The user may not be very clear in expressing their needs due to language issues, ambiguity in languages, use of jargons and much more. To reduce the impact of language on requirement specification, the customer journey could be used, i.e. users may specify their journeys in execution of a task, which is easier for other users to understand, which further intrinsically motivates them to participate. The challenge is to take advantage of the social interactions and counterbalancing it with the user diversity issues, this article proposes a new Requirement elicitation and prioritisation methods based on the following:

(a) Customer Journeys: This represents the activities customers have to undertake to finish the task. For instance, to pay for social insurance, he needs to login, select payment, make payment and print receipt. The rationale is that customer journeys help other users to better understand the viewpoint of other users and contribute in a better way to further refine the Journey.
(b) Crowdsourcing: The RE is conducted by involving crowd of users. To take into account their diversity, same crowd based RE is conducted with homogenous user segments (for instance, technology savvy users). The results of crowd based RE executed on different homogenous user segments are merged to yield single ranking of the requirements.
(c) Gamification: The games elements are used to motivate users to participate.

The proposed algorithm had been evaluated on the real data set of the social sector application used by leading Indian NGO. The NGO activists (who participated in the validation exercise) provided inputs about the software features and their priorities; which were then compared with the benchmark results; and results proved to be promising in social sector context. This chapter is structured as follows, Sect. 4.2 provides the theoretical background about SSRE, Sect. 4.3 provides the details about the proposed algorithm, Sect. 4.4 provides a hypothetical example, evaluation exercise performed in Sect. 4.5 and finally the chapter is concluded, and future work highlighted in Sect. 4.6.

4.2 Theoretical Background

Authors in [1] conducted a case study with multinational firms dealing with social sector software development and reported that the user segment diversity is the biggest challenge to conduct participatory RE. Single RE technique does not find its usefulness in social sector software development and blended RE techniques (intermix of traditional and crowd-based techniques) should be adapted as per characteristics of targeted user segments.

Author in [11] conducted a tertiary study of the literature and reported that primary and secondary crowdsourcing and gamification studies focusing on conducting RE in social sector context is too limited that make it difficult to build a theory; theory that could be useful for requirements engineers dealing with social sector software development. The area is still in its infancy and is yet to attract the researchers focus.

Limited primary work in the area is reported in studies [12–15] in the form of a vision and new methods but they have limited applicability to handle the social sector unique challenges especially diversity [1]. Although primary studies that focus on social sector requirements engineering provide a good basis for knowledge building, yet they are too limited that could help to formulate a unified viewpoint. These studies are silent on how diverse users (including elder people) could be involved in requirements engineering decision making activities and how their perspectives will be accounted for. There is now an agreement that crowd supported requirements engineering is the focus of researchers these days but the way for ensuring the equal representation of user groups is yet to be investigated. Table 4.1 provides a snapshot of the brief description of the limitation of the available literature of social sector RE.

4.3 Working Algorithm

The algorithm elicits the customer journeys from the users (i.e., the citizens), which are then subjected to further elaboration and prioritisation by initiating the social interactions between the users. Customer Journey (J) is the set of N values that represent the "N" journeys of the customers which represents their needs and benefits. The requirements engineering activities are initiated by the requirements engineering engine while the users are motivated to continuously participate in the activities for the richer social interactions through the gamification engine.

To manage the constraint of diversity of the user base especially on the basis of the ease of use of the technological resources, this algorithm is fragmented into two sub-algorithms (Algorithm 1 and Algorithm 2). The algorithm 1 is meant for the users with ease of use to the technological resources. This algorithm allows the users to do the following:

The user can do the following (Requirements Engineering Engine):

- Post their Journeys (Posting).
- Add Sub-journeys to the already posted Journeys (Sub-posting).
- Rate the posted Journeys using the scale of 1 to 3 (3 means high priority) (Rating).

Gamification engines provide points to the contributors, which are used to identify the fake users and to normalise the journey ratings in accordance with the contributor reputation (measured in terms of their accumulated points). The contributors are provided with the points as per their participation in the process. The

Table 4.1 RE studies focused on social sector domain

References	Focus	Research type	Diversity addressed	Limitation
[12]	Collaborative mass user participation in requirements engineering	New method	No	1. How the diverse users (especially less technology savvy users) can be involved in requirements engineering? 2. How the perspectives of different groups will be merged (if different mechanisms are used to reach different user groups)
[13]		Vision	No	
[14]	Motivational modelling to reduce socio-technical gap.	Vision	No	
[15]	Collaborative mass user participation in requirements engineering.	New method	No	

gamification engine allots the user points for each interaction with the crowdsourcing engine, as per the following rules:

Posting:	2
Sub posting:	2
Rating on other posts:	1
Ratings given by another user:	1 (positive rating) or −1 (negative rating)

The general idea is that the Journeys with many sub-journeys signify the high priority of the Journey, which is to be translated into software requirement. Each individual journey (or sub-journey) is rated by the users and points are automatically updated. The individual ratings are normalised according to the average number of user points. The priority of the main journey is finally calculated on the basis of the number of sub-journeys and average ratings received by the sub-journeys.

The Algorithm 2 is meant for the less technology savvy people (for instance, old people or illiterate users). This algorithm provides to the users with the list of customer journeys, that are formulated in the way, suitable for users with special needs. The users can provide sub-journeys and rate the journeys. The priority of each journey is then finally computed. Each customer journey (and sub-journeys) has two values of the priority given by P (P1, P2), with P1 computer from algorithm 1 and P2 computed from algorithm 2 (Table 4.2). The final priority is computed using Algorithm 3 (Table 4.3).

Table 4.2 Algorithm 1

Let J = Set of N customer Journeys.
1. [Posting]
Update point( ).
[Update point( ) is the function that updated points associated with individual user].
2. [Sub Posting]
Update point( ).
3. [Rating]
Update point( ).
4. [Calculate Priority]
For each J_i in J, $Rating_i$ = ratings(i).
[ratings(i) is the function that returns the sum of rating given by users to individual journey i].
$Rating_i = Rating_i / Avg\ (Points)$
[This helps to overcome the wrong ratings by the fake users or mistakes]
[This step normalises the rating of each journey with the average points of the contributors].
For each J_i in J, Number = child(J_i)
[This step calculated the number of sub journeys associated with the journey i].
Priority (J_i) = Number * ($\sum Rating_i/N$) , for i = 1 to N.
Create three structure of the Journeys (and sub journeys) and mark the priorities. Call it Tree 1.

Table 4.3 Algorithm 2

1. Formulate the customer journeys expressed by the citizens (in algorithm 1) into short meaningful journeys to be communicated to people with special needs.
2. Send the journey descriptions to the citizens (less technology savvy people), with the following rating scale:

Category ratings	1	2	3
Meaning	Agreed	Disagree	Branch

3. Citizens rates each journey using the scale of 1, 2 or 3. In case of 3, they write in natural language that what extra value could be provided by addressing the category (Branch Journey).
4. **[Calculate Priority]**
For each J_i in J, $Rating_i$ = ratings(i).
[ratings(i) is the function that returns the sum of rating given by users to individual journey i].
$Rating_i$ = $Rating_i$ / Avg (Points)
[This helps to overcome the wrong ratings by the fake users or mistakes]
[This step normalises the rating of each journey with the average points of the contributors].
For each J_i in J, Number = child(J_i)
[This step calculated the number of sub journeys associated with the journey i].
Priority (J_i) = Number * ($\sum Rating_i/N$) , for i = 1 to N.
5. Create three structure of the Journeys (and sub journeys) and mark the priorities. Call it Tree 2.
6. Call Algorithm Merge.

Table 4.4 Algorithm 3: Merge

1. Call Algorithm 1 to elicit and rank customer journeys. Call the prioritised set as Rank 1.
2. Call Algorithm 2 to rank customer journeys as represented by Rank 1. Call the prioritised set as Rank2.
3. Computer agreement between Rank 1 and Rank 2 using Kendall's correlation coefficient. If the result is:
   (a) If $\tau = 1$ (perfect correlation) then, finalise Ranking and exit.
   (b) If τ != 1 (No correlation) then, go to step 4.
4. Create Crowd graph G (V, E) by analysing Tree 1 and Tree 2, using the following rules. V is the set of vertices and E is the set of edges. Each journey is represented by the vertices of the graph G. Each vertex is labelled using the triplet (Name of Journey, Number of participants (N), Computed priority (P)). Following rules applies:
   (a) For each Journey J, create a node in the graph G.
      (i) **Calculate Normalised vectors.**
         If N (Tree1) > N (Tree 2) then ΔN = N(Tree1) – N(Tree2)
         Else ΔN = N(Tree2) – N(Tree1)
      (ii) **Computer final priority, using:**
If N (Tree1) > N (Tree 2)
P(J_i) = (Priority (J_i from Tree1) + **ΔN** * Priority (J_i from Tree2))/2.
         Else
P(J_i) = (**ΔN** * Priority (J_i from Tree1) + Priority (J_i from Tree2))/2.
   (b) Repeat step (a) for each Journey J.
7. Exit.

Finally, the priority computed by the two sub-algorithms (algorithm 1 and 2) are mapped into the single priority based on the weighted average, considering the difference in the number of participants, using algorithm 3 (Table 4.4).

4.4 Hypothetical Example

Consider a simple system that is represented by two customer journeys denoted by J0 and J1. Journey J1 is composed of three sub-journeys J2, J3 and J4, respectively. The calculation of priorities based on ratings given by two technology Savvy users is given in Table 4.5. User U1 and U2 are considered to be equally participative in RE process, having points of 10 each.

Now assume that there are three equally participative but less technology savvy users (for instance, senior citizens) denoted by U3, U4 and U5 with equal points 10. The Calculation of priorities is given in Table 4.6.

To calculate the final priority of two user journeys J0 and J1, Algorithm 3 is executed. For the hypothetical system, N (Tree1) < N (Tree 2) so ΔN = N(Tree2) – N(Tree1); ΔN = 1. The priority is computed using:

P(Ji) = (ΔN * Priority (Ji from Tree1) + Priority (Ji from Tree2))/2
P(J1) = (1 * 1.65 + 2.79)/2 = 2.22
For J0, there is an agreement between P(J0) and P(J0), so P(J0) = 0.3

4.5 Result Validation

The algorithm was validated by undertaking the following activities:

(a) The social improvement programmes launched by the NGO foundation were screened and one such programme called "System for transparent financial transitions of charity funds" was selected as the benchmark. This system allows the users to donate the money, check how their money is being spent in real time and overall financial status. Blockchains were used for ensuring auditability and transparency in the process. The rationale behind selection of this system as a benchmark case was that the system had been in use from past 1 year and had

Table 4.5 Customer Journey Priority calculation for Technology Savvy Users

Customer journey	Sum of ratings	Priority (Algorithm 1)
J0	6	6/20 = 0.3
J1	4	3 * (4/20 + 2/20 + 3/20 + 2/20) = 1.65
J2	2	
J3	3	
J4	2	

Table 4.6 Customer Journey Priority calculation for less Technology Savvy Users

Customer journey	Sum of ratings	Priority (Algorithm 1)
J0	8	14/30 = 0.3
J1	9	3 * (9/30 + 7/30 + 6/30 + 6/30) = 2.79
J2	7	
J3	6	
J4	6	

Table 4.7 Mapping of values as per type of transactions

Type a	Informational (Checking Financial spending)	Navigational (Navigating website)	Transactional (Donations)
	3	**1**	**5**
Rationale	Users have keen motivation to check financial spending	User is motivated but is just checking details	User is highly motivated as reflected by his action of donating money

been quite successful in the market. In other words, the NGO has the details about the contributors and their interactions with the system. These details are used to sample out the users for performing the validation of the algorithm. Only those users are sampled that had created account with the system and had expressed their consent to receive the NGO updates and communications. Contributor details helped to divide the sample into two categories—G1 (Group of users which are in the age group of 18–45 years) and G2 (Users with 60 plus age). Users interaction with the system helped to allot them initial points using the following formulae:

Point = Σ *aa* * *Number of interactions with the system, from 1 to M, where M is max number of interactions that user had.*

aa: Constant that takes values in accordance with the type of interactions user have. The values are mapped to the interactions as given by Table 4.7.

(b) The G1 and G2 were approached using e-mails and their consent for the participation in the research was taken, before they were involved in the process of elicitation and prioritisation. Google spreadsheets were shared between the G1 and G2 users for felicitating the social interaction. The sample includes 24 users in G1 and 24 in group G2. The overall process was scheduled for the duration of 3 h.

(c) Following rewards were mentioned to the users for their active participation, which were given on the basis of attainment of the points, as given in Table 4.8.

(d) Group G1 was asked to mention their journeys/sub-journeys and rate other journeys. The negotiator (two JAX foundation members) were continuously witnessing the entire interaction of 3 h. The negotiators manually calculated users points and reformulated the journeys descriptions. The priority was also calculated on the basis of the user points and ratings (algorithm 1). The journeys, sub-journeys and priority are given in Table 4.3.

Table 4.8 Mapping of values as per type of transactions

Points	Reward
50	Invitation to core member group of the foundation
100	Invitation to the elite group of the foundation

Table 4.9 Customer journeys and priorities

Customer journeys	Group G1 priority	Group G2 priority	Final priority
J1: The user should enter his membership number, select the contribution and navigate to see the documents and progress images	2	4	3
J2: The user enters his credentials and could see the various ongoing projects seeking donations	3	3	3
J3: User should see ongoing projects information based on the location, progress and contributions made	4	3	3.5
J4: User should enter either select any ongoing project for making donations or donate for generic funds Two sub-journeys include: **J4(a):** Possibility to get consolidated information about the ongoing projects **J4(b):** Fund donation	4	6	5
J5: User should be able to easily navigate through the site to get more information before deciding their associations	2	3	2.5

(e) Group G2 were presented with the Journeys as elicited from the group G1. They were asked to mention their journeys/sub-journeys and rate other journeys. The negotiators manually calculated users points and reformulated the journeys descriptions. The priority was also calculated on the basis of the user points and ratings (algorithm 2). The journeys, sub-journeys and priority are given in Table 4.3.

(f) The priority list calculated for group G1 and G2 were having the Kendall's correlation coefficient value (τ) of 0.78, meaning there is strong correlation between the two ratings. However, as it is not unity, so the algorithm 3 was executed to calculate the final priorities. The final priorities are mentioned in Table 4.9.

The final ranking yields the journeys in the order J3 = J4, J1 = J2, J5. The actual access to the benchmark system on the basis of the user traffic and their interactions corresponds to the fact that in reality, the requirements corresponding to these journeys are executed in the same order.

4.6 Conclusion and Future Work

The success of the technological solutions depends on their ability to meet customer needs, for which RE holds paramount importance. In the social sector context, it is very challenging to identify and rank the needs of diverse customers because it is hard to perform collective intelligence by aggregating the user perspectives which is distributed among the crowd. Blended RE techniques with different crowdsourcing platforms should be adapted as per user segment characteristics and perspectives brought from each segment participation should be merged in a rational way. This article presents one such approach where RE techniques are executed over a crowd of homogenous user segments, motivated using gamification and support by suitable crowdsourcing platforms to identify and rank customer journeys. The individual results are then merged resulting in single ranking of the customer journeys. Social interaction through customer journey helps to make perspectives of the user posting his needs, much visible to other users, which help to limit the chances of false positives. In future, it is expected that this technique could be scaled across different social networks and its efficiency validated across millions of citizens as users as a longitudinal study.

References

1. V. Gupta, Requirement Engineering Challenges for Social Sector Software Development: Insights from a Case Study, Digital Government: Research and Practice. (Under Review)
2. E.C. Groen, N. Seyff, R. Ali, F. Dalpiaz, J. Doerr, E. Guzman, M. Hosseini, J. Marco, M. Oriol, A. Perini, M. Stade, The crowd in requirements engineering: The landscape and challenges. IEEE Softw. **34**(2), 44–52 (2017)
3. J.A. Khan, L. Liu, L. Wen, R. Ali, Crowd intelligence in requirements engineering: current status and future directions, in *Requirements Engineering: Foundation for Software Quality. REFSQ 2019. Lecture Notes in Computer Science*, ed. by E. Knauss, M. Goedicke, vol. 11412, (Springer, Cham, 2019). https://doi.org/10.1007/978-3-030-15538-4_18
4. R. Sharma, A. Sureka, CRUISE: a platform for crowdsourcing requirements elicitation and evolution, in *2017 Tenth International Conference on Contemporary Computing (IC3)*, Noida (2017), pp. 1–7. https://doi.org/10.1109/IC3.2017.8284308
5. R. Snijders, F. Dalpiaz, S. Brinkkemper, M. Hosseini, R. Ali and A. Ozum, REfine: a gamified platform for participatory requirements engineering, in *2015 IEEE First International Workshop on Crowd-Based Requirements Engineering (CrowdRE)*, Ottawa, ON (2015), pp. 1–6. https://doi.org/10.1109/CrowdRE.2015.7367581
6. M.Z. Kolpondinos, M. Glinz, GARUSO: a gamification approach for involving stakeholders outside organizational reach in requirements engineering. Requir. Eng. **25**, 185–212 (2020). https://doi.org/10.1007/s00766-019-00314-z
7. R. Cursino, D. Ferreira, M. Lencastre, R. Fagundes, J. Pimentel, Gamification in requirements engineering: a systematic review, in *2018 11th International Conference on the Quality of Information and Communications Technology (QUATIC), Coimbra* (2018), pp. 119–125. https://doi.org/10.1109/QUATIC.2018.00025
8. M.Z. Huber Kolpondinos, M. Glinz, Behind points and levels — the influence of gamification algorithms on requirements prioritization, *in 2017 IEEE 25th International Requirements*

Engineering Conference (RE), Lisbon (2017), pp. 332–341. https://doi.org/10.1109/RE.2017.59
9. F.M. Kifetew, et al., Gamifying collaborative prioritization: does pointsification work?, in *2017 IEEE 25th International Requirements Engineering Conference (RE)*, Lisbon (2017), pp. 322–331. https://doi.org/10.1109/RE.2017.66
10. F. Kifetew, D. Munante, A. Perini, A. Susi, A. Siena, P. Busetta, DMGame: a gamified collaborative requirements prioritisation tool, in *2017 IEEE 25th International Requirements Engineering Conference (RE)*, Lisbon (2017), pp. 468–469. https://doi.org/10.1109/RE.2017.46
11. V. Gupta, Crowdsourcing and gamification in requirement engineering of social sector applications: a tertiary study. IEEE Trans. Technol. Soc. (Under review)
12. D. Renzel, et al., Requirements bazaar: social requirements engineering for community-driven innovation, in *Proceedings of IEEE RE* (2013), pp. 326–317
13. T. Johann, et al., Democratic mass participation of users in requirements engineering? in *Proceedings of IEEE RE* (2015), pp. 256–261
14. G. Ruhe, et al., The vision: requirements engineering in society, in *Proceedings of IEEE RE 2017*, Sept 2017, pp. 478–479
15. R. Burrows, et al., Motivational modelling in software for homelessness: lessons from an industrial study, In *Proceedings of RE* (2019), pp. 298–307

Chapter 5
Implications for Social Innovation Stakeholders

Abstract The social innovation requires active involvement of all open innovation ecosystem elements including citizens. The social innovation actual impacts are evident if the technology solves the real social needs and is suitable to be adopted in the working context of the citizens. This chapter highlights the various implications on innovation ecosystem elements—researchers, academia, Government, funding agencies, social entrepreneurs.

Keywords Social innovation · Technologically infused social innovation · Researchers · Funding agencies · Social entrepreneurs · Government · Academia

5.1 Introduction

The social sector is briefly introduced in Chap. 1 and various challenges and state of technological solutions available to undertake RE in social sector context were presented in Chaps. 2–4. The social sector RE is challenging to be performed and needs strong collaboration across software engineering research boundaries. Social innovation requires the active support of different stakeholders-researchers, academia, Government, funding agencies, social entrepreneurs, etc. The open innovation will help to foster two-way knowledge sharing between the stakeholders leading to the commercialisation of interdisciplinary solutions with the ability to meet social needs. Social innovation needs better understanding of the social domain, products based on rigorous solutions solving social problems, continuous support of the citizens and continuous measurement of innovation potential.

The citizens' support is crucial because it is ultimately them who are going to be impacted by the innovation and success of social innovation depends on how easily it has been adopted by them. However, a good technology may sometimes fail to

---

Some of the text of this chapter appears in ''Varun Gupta (2021) Requirement Engineering Challenges for Social Sector Software Development: Insights from Multiple Case Studies. Digit. Gov.: Res. Pract. https://doi.org/10.1145/3479982. © 2021 Copyright held by the owner/author(s)''.

V. Gupta, *Requirements Engineering for Social Sector Software Applications*, https://doi.org/10.1007/978-3-030-83549-1_5

bring useful results because citizens are unable to adopt them owing to numerous factors like complexity, costs and ease of use. It is important to ensure that the real benefits of technology should be visible overcoming the adoption related issues by active support of all stakeholders.

The social innovation requires the active support of different stakeholders-researchers, academia, Government, funding agencies, social entrepreneurs, etc. The open innovation will help to foster two-way knowledge sharing between the stakeholders leading to the commercialisation of interdisciplinary solution with the ability to meet social needs. The social innovation needs better understanding of social domain, products based on rigorous solutions solving social problems, continuous support of the citizens and continuous measurement of innovation potential.

The citizens support is crucial because it is ultimately them who are going to be impacted by the innovation and success of social innovation depends on how easily it had been adopted by them. However, a good technology may sometimes fail to bring useful results because citizens are unable to adopt them owing to numerous factors like complexity, costs and ease of use. It is important to ensure that the real benefits of technology should be visible overcoming the adoption related issues by active support of all stakeholders.

The overall objective is to explore the market needs by involving the diverse citizens across geographical boundaries, making the requirements engineering a co-creation process leading to the product that can be easily adopted by the citizens. The support of all innovation ecosystem elements is must for the requirements engineering.

5.2 Implications for Researchers

The researchers have the competencies to provide rigorous research solutions with the ability to solve societal problems. These research solutions once commercialised as a product could be a very powerful tool to reform the society and bring overall development. The researchers should also try to take entrepreneurial risks with the support of incubators, funding agencies and other innovation ecosystem elements to commercialise their innovations. There is a need to bring proximities between the different departments, research centres and external social innovation stakeholders to result in interdisciplinary solutions to social problems. They should investigate how the gaps between the citizens' competencies in using the technology and the technological complexity could be bridged.

Furthermore, apart from these management issues, researchers need to investigate the ways by which the diverse citizens could be involved in the process of exploring the problem domain to find a solution with good market fit. The challenge as outlined in Chap. 3 is that computationally advanced requirements engineering techniques may not work when dealing with the special groups, especially ageing people, who are not technology savvy. How can the requirements engineers be able to identify the distributed scattered information across geographically located citizens and then converge it into useful system related information to drive release planning, the issue that remains unaddressed as of now?

5.3 Implications for Funding Agencies

There is a growing focus of the federal governments and other funding agencies to support the entrepreneurial activities in social sector. The reason is that social entrepreneurial activities could be the powerful way of fostering social innovation leading to strong social impacts. For instance, European Union (EU)[1] has been funding various social entrepreneurial activities as well as research projects with social impacts. The real innovation of the project is visible after a long period of time, meaning that if the focus is to promote the projects with good social impact, then efforts should be made to accurately predict the social impact of the project proposal. The efforts should be made to provide the resources to the social entrepreneurs by providing access to the professional networks with which the funding agency has close proximity. This will help to leverage the expertise of the diverse group across various sectors and competencies.

There is a balance to be maintained whether the funds are to be allocated to projects related to more technical related aspects (for instance, requirements engineering or building requirements engineering systems for ageing people) or it should more focus on management related aspects (for instance, field studies to identify social problems or solutions for social problems). The right solution expected will be the focus on the interdisciplinary projects with both technical as well as management projects; the issue that could be well addressed if project proposals should only be diverse and interdisciplinary teams from a group of research centres, universities and industries as a single consortium.

5.4 Implications for Government

The social sector reforms had traditionally been the responsibility of the governments. Public sector institutions are held responsible for providing their services for removing social problems in order to benefit the community. Government institutions undertake social innovation through the adoption of technology for which they outsource their needs to third party software development companies (including public undertakings). The software firms are having best expertise in undertaking technical work, but they lack the deeper understanding of the societal challenges (expect those that they observe in daily lives).

The accessibility of the community will be a challenge for the requirement analysts. High quality software will provide value to the citizens that indirectly will be valuable to the governments in meeting their objectives. The ability of the solution to meet the needs of the community and getting accepted by them will require a requirement analyst to be familiar with social problems. For this to happen, the role of government is quite important. The government support will not only help

[1] https://ec.europa.eu/social/main.jsp?catId=952&intPageId=2914&langId=en

software firms to develop good quality software, but it will provide the research community with access to diverse experiences that will drive future research resulting in rigorous requirements engineering solutions.

However, the emergence of social entrepreneurs has been one of the supports to the government in terms of sharing responsibilities for social good. The role of Government thus goes beyond providing good access to the masses for collecting requirements to providing support to the social entrepreneurs by easing the regulations, tax holidays, access to international markets, funding and other required resources. The overall idea is that social entrepreneurs could be best utilised in either of three ways—implementing the technologically induced social innovation in society by acting as service providers, as a promoter of technology among the masses to make its adoption easier and to be the reason for technological social innovation as technology owners.

5.5 Implications for Academia

Academia can play an important role in promoting social innovation by providing their support for establishing the social entrepreneurs and contributing by commercialising the research based social impact projects. Their support for promoting social entrepreneurs is important as they could support open innovation by partnering with the start-ups thereby promoting the two-way knowledge sharing, leading to commercialisation of products with social impact. The academia and start-up partnership could be the best way to have open innovation [1]. The academia's third mission is to go beyond their "boundaries" and provide support for societal reforms. Social innovation could be fostered with academic support in the form of commercialisation of "In-house" research, support for entrepreneurial activities, partnering with third parties for solving societal problems, etc.

Academia should try to create synergies between their research units to provide solutions for doing requirements engineering with special groups; the outcome of which should help the requirement analysts to explore the problem domain of the social sector with ease. The units that could be useful in this task could be those dealing with ageing people research, marketing department (they are closest with the people), natural language processing, software engineering, social sciences and management.

5.6 Implications for Social Entrepreneurs

Social entrepreneurship is considered to be an array of wider range of activities including- enterprising individuals devoted to making a difference; social purpose business ventures dedicated to adding for-profit motivations to the non-profit sector; new types of philanthropists supporting venture capital-like 'investment' portfolios;

and non-profit organisations that are reinventing themselves by drawing on lessons learned from the business world [2].

The social entrepreneurs establish their business activities by getting motivated to make social contributions rather than by the profits to be earned. The major challenges faced by social entrepreneurs include those related to policymaking, legal aspects, institutional and operational support, social, educational and cultural awareness of the field and its ecosystem [3].

The start-ups have higher failure rates, and their community faces unique challenges especially limited resources like human resources and financial resources [4–12]. The social start-ups being motivated with social contributions operate being funded by charities, donations, federal government support, etc. as the objective is not to earn profits.

The social entrepreneurs have a great contribution to make in society either by launching new products and services or by acting as service providers of the government technological innovations to the citizens. In either case, the support of the Government is needed for these entrepreneurs to survive. These entrepreneurs could take advantage of crowdsourcing, open-source technology, crowd funding and available funding support (for instance, the funding schemes of European Union) to support their business operations within limited resources and have sustainable growth. However, as their social innovation reaches the society and actual impact becomes visible, the scalable growth could result in increased funding support by the citizens. The business growth and social impact are interrelated, with one impacting the other equally.

The social entrepreneurs could have specialisation in any domain-technical, management, medical sciences, etc. Their expertise strongly determines the tactical actions they choose to operate their social businesses. A good expertise in social management is a strong competency as they have a rich understanding of underlying challenges and issues. However, they need to collaborate with technical people to share their knowledge about society to help them undertake requirements engineering followed by software development. Similarly, a good technical background is a key competency but support of those who are knowledgeable in social sciences is a must as they hold the key to unlocking market information. This strongly signifies the need for diverse interdisciplinary teams in social start-ups/enterprises.

5.7 Conclusion and Future Work

The diverse team competencies in social enterprises are a key to success in business owing to the fact that people don't buy technologies, they buy solutions. In the social sector, the focus being more on social improvement, the pricing of the social product is either assumed by the federal government or covered by funds; finally, the citizens either consume services free or pay marginal costs as service charges. The product development costs, and other transactional costs need to be kept lower, the issue that grounds on accurate as well as cost aware RE. The rich knowledge sharing between stakeholders will be a key to undertake RE with optimal resources and efforts.

References

1. V. Gupta, J.M. Fernandez-Crehuet, D. Milewski, Academic-startup partnerships to creating mutual value, in *IEEE Engineering Management Review*. https://doi.org/10.1109/EMR.2021.3065276
2. J. Mair, J. Robinson, K. Hockerts (eds.), *Social Entrepreneurship* (Palgrave Macmillan, New York, 2006), p. 3
3. A. Seda, M. Ismail, Challenges Facing Social Entrepreneurship. Review of Economics and Political Science (2019)
4. R. Chanin, L. Pompermaier, K. Fraga, A. Sales, R. Prikladnicki, Applying customer development for software requirements in a startup development program, in *Proceedings of the 2017 IEEE/ACM 1st International Workshop on Software Engineering for Startups (SoftStart)*, Buenos Aires, Argentina, 21 May 2017, pp. 2–5
5. C. Giardino, N. Paternoster, M. Unterkalmsteiner, T. Gorschek, P. Abrahamsson, Software development in startup companies: The greenfield startup model. IEEE Trans. Softw. Eng. **42**, 585–604 (2016)
6. M. Unterkalmsteiner, P. Abrahamsson, X. Wang, A. Nguyen-Duc, S. Shah, S.S. Bajwa, H. Edison, Software startups—a research agenda. e-Inform. Softw. Eng. J. **10**, 89–123 (2016)
7. C. Alves, S. Pereira, J. Castro, A study in market-driven requirements engineering, in *Proceedings of the 9th Workshop on Requirements Engineering (WER '06)*, Rio de Janeiro, Brazil, 13–14 July 2006
8. E. Klotins, M. Unterkalmsteiner, T. Gorschek, Software engineering knowledge areas in startup companies: a mapping study, in *Proceedings of the International Conference of Software Business*, Braga, Portugal, 10–12 June 2015; pp. 245–257
9. V. Gupta, J.M. Fernandez-Crehuet, T. Hanne, R. Telesko, Requirements engineering in software startups: a systematic mapping study. Appl. Sci. **10**, 6125 (2020). https://doi.org/10.3390/app10176125
10. V. Gupta, J.M. Fernandez-Crehuet, C. Gupta, T. Hanne, Freelancing models for fostering innovation and problem solving in software startups: an empirical comparative study. Sustainability **12**, 10106 (2020). https://doi.org/10.3390/su122310106
11. V. Gupta, J.M. Fernandez-Crehuet, T. Hanne, Freelancers in the software development process: a systematic mapping study. PRO **8**, 1215 (2020)
12. V. Gupta, J.M. Fernandez-Crehuet, T. Hanne, Fostering continuous value proposition innovation through freelancer involvement in software startups: insights from multiple case studies. Sustainability **12**, 8922 (2020). https://doi.org/10.3390/su12218922

Index

V. Gupta, *Requirements Engineering for Social Sector Software Applications*, https://doi.org/10.1007/978-3-030-83549-1

Printed in the United States
by Baker & Taylor Publisher Services